THE EVOLUTIONARY ORIGIN OF HUMAN BEHAVIOR

How Play and Evolution Carried Us
from Our Reptile Predecessors
To
The Story-Tellers We Are

By
Keith C. M. Glegg

iUniverse, Inc.
New York   Bloomington

The Evolutionary Origin of Human Behavior
How Play and Evolution Carried Us from Our Reptile
Predecessors to The Storytellers We Are

iUniverse books may be ordered through booksellers or by contacting:

iUniverse
1663 Liberty Drive
Bloomington, IN 47403
www.iuniverse.com
1-800-Authors (1-800-288-4677)

Because of the dynamic nature of the Internet, any Web addresses or links contained in this book may have changed since publication and may no longer be valid.

ISBN: 978-1-4401-1806-7 (pbk)
ISBN: 978-1-4401-1808-1 (cloth)
ISBN: 978-1-4401-1807-4 (ebk)

Library of Congress Control Number: 2009923907

Printed in the United States of America

iUniverse rev. date: 3/11/2009

This book is the reply to my brother Ron and his wife Grada who, during a recent get-together, recalled a manuscript of mine—something ... about play and evolution—that they had seen some twenty years ago, and asked what had become of it. Thanks to you both for reminding me.

I also want to say Thank You to Samantha, my PSA, Matt Abdon, my book designer, and Erin, my cover designer, for their help and dedication to bringing my book to life.

CONTENTS

ILLUSTRATIONS

INTRODUCTION

There is a quite natural question that one tends to ask about books like this, and you will almost certainly ask it, especially as you read further along. The question is: How did such a book come to be written? The answer is that, about twenty five years ago, I started assembling some thoughts for an essay entitled *"The Evolutionary Origin of Play,"* but, as I continued to explain the nature of play to myself, its possible significance began expanding to include a host of other interests that had beguiled me for at least the previous fifteen years. So what started as just a short essay, of which portions remain in Chapters II and V, grew into the book you are reading.

These "other interests" included questions concerning how little was really known about behaviors such as sleep and storytelling, which consume very large parts of our lives. Running parallel to these had been an interest in the almost incredible sweep of Darwin's theory of evolution. As a result of all this, although physics and engineering have supported my work, subjects like play, sleep, language, and evolution provided my real intellectual challenges. So what you will be reading is the result of the coming together of all these interests, and others like them, within a single framework which it seemed possible to erect. What became apparent to me was that it would be possible, beginning with an explanation of the evolutionary origin of play, to assemble an explanation of the evolutionary origin of *human behavior.*

Evidently, to have any hope of knowing where to aim the explanation, and when to end, requires some fairly precise idea of what is the defining feature of "human behavior." As you will see, the defining feature of evolution's arrival at "human behavior" is taken to be the emergence of the ability to *narrate,* to tell stories. As a consequence, in going from the emergence of play to that of narration, a large part of the explanation is concerned with tracing out the evolution of brain—from reptile to human brain.

Of course, not all of this is new, since neurobiology had been keenly aware that various *parts* of the brain—expansion of the upper brain stem; the limbic system; the cortex—must have emerged at different times, as essentially evolutionary developments. But what was missing is a *single evolutionary theory of human behavior,* accompanied by that of a brain, in which all the parts fit together naturally, including at least some suggestion of how they would have to work in order to have been crucial in human evolution, and the evolving *external behavior* that would have been needed to sustain human survival.

The situation in the evolutionary theory of human behavior was not unlike that which existed in *geology,* prior to the development of the theory of continental drift and plate tectonics. Before this development, geology was an essentially descriptive pursuit which, although it recognized the evidence of evolutionary processes in the rocks and sands that are its subjects, it had no *single,* consistent way of connecting the *present overall behavior* of Earth's surface with some evolving version of its *past overall behavior.*

So what geology needed was exactly what the theory of continental drift gave it, that is, a single evolutionary theory of how the surface of Earth came to look and behave the way it does. At this point, it became possible to bring the behavior of continents, mountains, earthquakes, volcanoes, oceans and much else within a single evolutionary theory that points to processes to which physics and chemistry could be applied directly. Thus, in a very real sense, what Charles Darwin did for biology with his theory of evolution in the nineteenth century, Alfred Wegener did for geology with his theory of continental drift in the twentieth.

What the evolutionary origin of human behavior needs is an analogous theory. For while neurobiology, for instance, recognizes the evidence of evolutionary processes in the emergence and development of the *parts* of the nervous system, there is no way of connecting our present behavior, or that of other animals, with some evolving version of past behavior. Just as in the case of pre-drift geology, in which the evidence for evolution of the parts was recognized, without there being any single evolutionary theory of overall behavior, neurobiology recognizes the evolution of the parts, especially the parts of a brain, without there being any single evolutionary theory of overall animal behavior as should form the basis of a biology-based, which is to say, an evolutionary theory of the origin of human behavior.

While it would be beyond hubris to say that the missing analogous theory for the origin of human behavior is what is provided here, it is perhaps fair to say that a sketch is given of what such a theory would look like. For what the sketch tries to show is how evolution might answer the nagging questions about the nature of the relationship between play, and sleep, and language, and music, and storytelling, and art, and time, and schizophrenia, and science, which are all to be found in human behavior. These are human behavior's oceans and mountains, its earthquakes and volcanoes, and hence the proper concerns of an evolutionary theory of its origin. So what the sketch shows is how it might be possible to link these within a *single* evolutionary framework, which reaches all the way from play to schizophrenia.

Since all these expressions of brain must be interconnected, it would doubtless be possible to start with any one of them as the place from which to build an evolutionary theory of brain-related human behavior. However, I have started with play, partly because of its sheer fascination; partly because it has attracted a large amount of devoted and carefully recorded observation on which it is possible to draw in order to get started; but, as I indicated earlier, mainly because of an unaccountable accident of many years ago. However, this turns out to have been a most rewarding accident. Since, as you will see, when play is placed in its full evolutionary context, it appears to provide the bridge between the essentially instinctive brains of reptiles, and the much more powerful brains of birds and mammals. Indeed, it is in the unique evolutionary shade of play that creatures appear to have made the first little steps toward a brain that can "understand."

You will find, as you read on that, although, in most places, the book is "scientific" in what is being discussed, it is not *just* a "scientific book." And that statement, even risky as it is, is worth making, since it leads to one of the book's advantages. For there are a number of places where I describe ways of seeing things in which they have never been seen before, and, in just a "scientific book," one would not share with the reader the excitement of finding a new way of seeing something. But, having burned my bridges, I have allowed myself the unhindered pleasures of the other side of the river, resorting, on occasion, to even joyous signs of exclamation!

The book goes even further beyond just science, to the occasional admission of humor, since this is another way of sharing feelings with the reader. This time, though, it is sharing the relief that sometimes comes at three in the morning, during a running dialogue with a cathode-ray tube, as it looks out through its multitude of little colored eyes. However, these occasional expressions of humorous relief also affirm my belief that one doesn't have to be unrelentingly grim to be serious—even very serious.

But, although this is not just a "scientific book" it is not an "unscientific book" either; and this also needs to be said, not to be needlessly foolhardy, and so lose all hope of finding a good and willing publisher. Indeed, I have included a considerable number of quotes from, and references to books and journals which anchor the explanation in what are generally taken to be bays having secure scientific bottoms, known from repeated soundings—not always by the same people—yielding much the same results.

Even here, though, and especially in the quotes, I have tried to make use of more readily available material, although this has not always been possible. It could be that this last failing is a reflection of my having, through the accident of my second place of work, too easy access to a national science library of immense proportions, staffed with truly helpful people; but it might, more properly, be simply a measure of the real difficulty of some of the subjects treated. Of course, I thank the numerous authors whose work has been quoted, and absolve them of any nonsense related to its use.

In this connection, it is worth adding that the language and general treatment is such that any reasonably curious person, albeit with a fair amount of effort in some places, should be able to get safely from one end of the book to the other without going to references, big dictionaries and the like. Chapter I, "Evolutionary Explanation," is a brief treatment of aspects of evolutionary theory which is intended to assist with this degree of self-containedness.

But the remaining chapters of the book are neither so straightforward nor conventional. So, in some earlier versions of this introduction (of which there have been many), there were summaries of each chapter. However, partly in the interest of limiting the length of an already overly long Introduction, this has been abandoned; but the absence of the summaries is also intended to allow you to better enjoy the surprises that, I'm quite certain you will experience, as you read along. If your curiosity is still not satisfied, then a more careful reading than usual of the CONTENTS should help. But therein lies the temptation to go directly to some favorite topic, which will likely be unrewarding because of the extent to which the sense of each chapter depends on that of the one preceding it.

Finally, there's the book's style—every book must have a "style." As you might have guessed, from just this Introduction, the book's style goes back to my enjoyable stints as lecturer in two universities. So the Chapters tend to be make-believe lectures to a group interested enough to ask un-answerable questions. As a result, the book, for better or worse, "sounds" like me. What is missing, though, is what *I* would have learned in the back-and-forth of real lectures, and which is never to be had from this kind of literary *sens unique*.

CHAPTER I
EVOLUTIONARY EXPLANATION

Explanation

All we need do is look around us to be convinced that humanity and its works are special. Indeed, so compelling is the evidence that we might easily come to believe that we are "fully detached" from the rest of the living world. But, beginning little more than a hundred years ago (now fifty years more than that), attention began to be directed toward even more compelling evidence of our *attachment to,* rather than our *detachment from* the world of the living, as it stretches back and converges toward some, as yet, unknown beginning. What the evidence suggests, as it continues to accumulate, is that we are, for instance, in a measurable biological sense, less different from the chimpanzee than it is from some kinds of monkeys. What the evidence increasingly suggests is that we, like all other members of the living world which exist and ever have existed, evolved, generation after generation, from previous forms.

But there are gaps in the evidence, some of which can be filled easily, and some less so. One of the latter relates to the question of how an "animal" could come, by this process of evolution, to what we experience as "selfness" or as "understanding." Is it possible to explain away such gaps simply as natural developments in a long process of evolution? In trying to see what the answer might be, it is useful to begin by looking at some of the factors that contribute to valuable and credible explanation.

Above all else, really valuable explanation must bring together a number of what had previously been taken to be *un*related phenomena. The reason for this is simply that any particular phenomenon, by itself, is usually "explainable" in a number—perhaps an unlimited number—of different ways. So it is only on expanding the range of the "separate" phenomena explained by a *single* explanation that we begin to reduce the number of possible explanations and demonstrate the power of any particular one.

Of course, no imaginable explanation will explain everything, but it is reasonable to expect of any explanation that we should not be able to present it with some phenomenon which would require it to contradict itself in order to sustain explanation.

Another requirement that might reasonably be placed on any explanation is that it be so constructed as to be "falsifiable." What this means is that an acceptable explanation will lead to some conclusion which a demonstration or experiment can either support or not support. Only explanations that have the capacity to bring on themselves and *survive* this kind of filtering by demonstration can be extended, with any assurance, to still further explanation.

Looked at in this context, evolutionary explanation, both for what it has explained and how it achieves explanation, appears as one of the great triumphs of the human brain. For, in one grand sweep, it tells us of a relation between all the biological phenomena that are to be found in every living entity that has ever existed and which will ever exist, at least on Earth. One of the ways in which it accomplishes this is by bringing sharply into focus the issue of survival.

SURVIVAL

For any living thing to have a continuing presence in the evolutionary world, that is, to survive, it must be able to meet challenges to both its physical and its physiological integrity.

By "physical integrity" I mean the ability of the creature to hold together physically. It needs to be able to resist being arbitrarily torn apart by external occurrences such as small changes in temperature, currents in water or air, or the destructive efforts of predators.

By "physiological integrity" I mean its capacity to resist simply falling apart from starvation. This entails primarily ensuring sufficient intake of food and adequate means for getting rid of waste.

REPRODUCTION AND EXTINCTION

But, in order that a creature might have lasting evolutionary significance, something else needs to happen beyond this self-directed survival activity, which is going on continually between the creature itself and its immediate environment. Out of this self-directed activity in the here-and-now must come some reproductive process which yields a reasonably faithful replica of the creature, so as to form a next generation. And since this concerns evolutionary explanation, I need to remind myself, at least, that such explanation cannot include the notion that creatures act in ways, non-self-directed ways, which favor reproduction as such. The nature of the evolutionary process requires that a creature act simply to benefit itself, that is, act in its self-interest. The interactions between the creature and its environment then engender and support reproductive processes or they don't.

Evidently, not every creature will continue to enjoy a relation with its environment that supports continued reproduction. In the case of those that do, we say the environment "favors" the creature, by a process of "natural selection" (to use Charles Darwin's phrase), for as long as this continues. When the efforts of the creature to sustain its own integrity no longer yield reproductive processes as well, there is finally no next creature, and the line of creatures, going back some number of generations from the last, is said to have become "extinct." This is simply another outcome of the relation between the creature and its environment, also mediated by natural selection.

SPECIES

No process of reproduction or copying will produce perfect replicas forever. In particular, the process underlying replicas of living creatures is subject to small errors, mutations, which tend to persist in successive replicas. These mutations can, on occasion, alter the relation between the creature and its environment in such a way as to alter the reproductive process itself. When the reproductive process is favorably affected, we say the mutation is favored by natural selection. Another way of saying the same thing is that the "fitness" of the creature is increased by the mutation.

Evidently, the combination of a more or less steady stream of mutations with the endless filtering provided by the environment in the form of natural selection would lead to the emergence, in successive generations, of particular groups of creatures which enjoy increasingly special relationships to the environments in which they live. When such a group becomes so specialized and distinct from those around it that either no mating outside the group occurs, or such mating, if it occurs, no longer leads to successive generations, we refer to the new distinct group as a "species."

The process that leads to such a complete separating-out can hardly be very rare, since there are estimated to be between ten and twenty million existing species. Humans constitute one of these species, and my object in this book is to explain some of the processes that have led to *our* separating out as the distinct species we have become.

GENES

As it happened, it was possible for the remarkably brilliant Charles Darwin to assemble the first comprehensive statement of the nature of biological evolution without an awareness of what has since become its cornerstone. Darwin's observations and those of others, on both plants and animals, on which he based his revolutionary conclusions about the nature of evolution, were almost all made at or near the level of the unaided human senses. What has become clear, however, since his early efforts in about 1840, is that what he concluded from observations at this large scale can be traced to the workings of sub-microscopic chemical groupings that are known as "genes," these workings being referred to as the "genetic" process.

It is now known that the form and much of the behavior of all living entities are determined by the action of individual genes or groups of genes, and that these form parts of long chemical chains in which they are held together in patterns that are characteristic of the particular living entities of which they form a part. It might seem surprising that massive creatures like elephants and whales, and even ourselves, should have all their form and much of their behavior determined by such sub-microscopic strings of genes, but this is now well known to be so. It is therefore not uncommon to see references to genes being "expressed" in various features of living entities.

It is these genes, then, and the long chemical chains that link them, which carry the information that allows one generation of a species to produce offspring of the same species. So when reference is made to "mutations," it is within genes that the chemical mutations occur which account for the changes that can take place within species, from one generation to the next. And it is these genetic mutations, as they continue to occur, while their effects continue to be filtered by natural selection, which lead eventually to new species.

Evidently, the chemical assemblies that make up genes must be relatively stable in order to convey, from one generation of a species to the next, essentially the same information about its members. It is therefore reasonable to wonder how this stability can be reconciled with the mutations that seem to be necessary to give rise to new species. One answer to this is that, even though genes are stable in the presence of the most frequently occurring forces which would tend to disrupt them, such as the continual bombardment due to thermal agitation in the media in which genes reside, there is, at the surface of the earth, a steady flow of particles which are even smaller than genes, that originate both inside and outside the earth, and which have sufficient energy to easily disrupt a gene with which they might collide. Although such collisions might occur infrequently in the life of any particular creature, they will occur more frequently in the species as a whole, and it requires only one member of a species to suffer a genetic mutation in order to set in motion a possible movement toward a new species. It is therefore reasonable to suppose that there is a steady stream of creatures in which mutations take place. Natural selection will suppress most of the mutations as expressed in the creatures which bear them, but some of the creatures carrying mutations will survive. As the process continues, new species will eventually appear. In this way, the necessary stability of genes is overcome infrequently but steadily, thus giving rise to the whole progression that we call evolution.

This, as many readers will know, is hardly even a sketch of what genes are, not to mention how they perform their marvels. But it will suffice for what I am setting out to do, for much the same reason that Darwin was able to do without genes altogether. The reason, of course, is what one might call the "level of disaggregation" of living creatures at which the discussion will proceed. I shall be proceeding mainly at the level of the whole creature, and two levels down, such as a brain and parts of a brain. As you will see, when proceeding in this way, it is convenient to be able to refer to genes at about the level at which they have just been described, but going any further here with their description would contribute little to what I am setting out to do.

INSTINCTIVE BEHAVIOR

Since the young of many creatures emerge from eggs that hatch in isolation from their parents, and, in spite of this total isolation at birth, immediately execute behavior typical of their species, it is clear that such behavior must be part of the development of the creature that is determined by its genes. Such behavior is what we usually refer to as due to "instinct," that is, as "instinctive" behavior. Since instinctive behavior is determined by genes, such behavior must be passed on from parents to children, that is, must be "hereditary." However, in view of the possibility of a mutation occurring in a parent which is not expressed in that parent itself, but only in its children, "hereditary" behavior can only imply similarity, with some room for shifts due to parental mutations.

Of course, it is not only behavior that is executed immediately after birth that is instinctive, but, as a creature has a chance to mix with other members of its species, especially older members, and its behavior becomes more and more complex, it becomes increasingly difficult to identify, with assurance, which segments of the creature's behavior are instinctive and which are not. Indeed, one might wonder whether there are *any* segments of behavior that are *not* instinctive.

A simple way of addressing this is to go directly to behavior in ourselves, in which we find components that are not hereditary. One such is language, since we quite evidently do not have to speak the same language(s) that our parents speak. If there is a hereditary component in language (and as I shall show in Chapter VI, there is) it certainly does not control language at the level of the "dictionary" we use—the child of ancestors who speak only Chinese might speak only French. So there is clearly, in us, a vast domain of behavior that is not hereditary, and that is acquired from contact with, and by *copying* the behavior of other members of our species, which might, but need not include ancestors.

Of course, we also carry out instinctive behavior, since we, like all other mammals, do not need to be shown how to suck, for instance, just after birth. It is therefore clear that some process of evolution must have occurred that would lead to our incorporating, *in the same brain,* hereditary programs that control *instinctive* behavior, on the one hand, and hereditary programs that allow us to acquire behavior by *copying,* on the other. Much of this book is devoted to seeing what this process might have been.

COPYING

It is important to notice that whether we refer to acquiring programs of instinctive behavior by means of *genes,* or of *non-*instinctive behavior by *copying as imitation*, we are, in *both* cases, witnessing *the power of copying,* since the genetic process is, beyond everything else, *copying.* Indeed, if one had to identify the most basic process at work in biology, it would quite properly be *copying.* What you will find, then, is that much of what is talked about in this book has to do with the variety of ways in which a brain becomes able to acquire and be controlled by processes of copying.

STRESS

We can come on a different way of viewing the evolutionary process by noticing that every creature, to survive, must be able to perform behavioral routines that allow it to continue satisfying its needs in the presence of a changing environment. But, although this must generally be the case, there will inevitably be situations in which, although a creature survives, there exists a distinct gap between the behavior that it *would have had* to execute in order to satisfy its needs *fully* and that which all the behavioral routines available to it *allow.* This gap, between behavior that would meet needs *fully* and what can *actually* be performed, sets up in a creature the condition we call "stress." Thus, stress is simply a reflection of the inevitable lack of a "perfect fit" between the behavioral endowment of a species and its environment.

But even if, over long periods in a creature's life, there could be a "perfect fit" in some average sense, there would still be short periods of heightened stress, such as those associated with recurring attacks of hunger, since there are no behavioral routines that can satisfy hunger *permanently.* Consequently, as the intensity of stress rises and falls, it can be viewed as "driving" much of a creature's short-term behavior, and as being able to change its short-term biological state.

It is therefore reasonable to say that surviving species are those that have evolved so as to limit the stress that their members experience, and so we can expect to find a variety of ways in which mutations and natural selection have combined to limit or eliminate stress in them. As we will see, *sleep* is one of these eliminators of stress, and it will serve to show how intricate a process for eliminating stress can become.

Of course, every surviving species is exposed to the possibility of extinction due to changes in its environment which would subject its members to a new and sufficiently severe condition of stress that cannot be reduced or eliminated by either existing behavioral routines or new routines that can be acquired soon enough to avoid extinction. In a related way, mutations in individual members of a species can lead either to the extinction of these members due to increased stress, or to conditions of reduced stress. In the latter case, the members and their offspring survive, and could ultimately lead, by the extension of such a process of mutation combined with the filtering of natural selection, to a new species in which there is a new kind of "non-perfect fit" between its behavioral endowment and its environment.

STEPS IN EVOLUTION

The next point I shall raise relates to the necessarily extremely small steps by which, in my view, evolution must proceed. That it should proceed in steps at all is clear enough since, being a biological process, it must depend, ultimately, on differences between various chemical substances. But, as we know, various chemical substances differ from one another in steps, simply because the parts that make them up are discrete atoms drawn from a limited set of about a hundred different elements. For there to be no steps, the chemical substances that form genes would have to be made up of parts drawn from some kind of continuous soup of possibilities; but, so far as we know, the world is not constituted in this way. So steps we shall always have. But why should there be "extremely small" steps?

Here we don't seem to have the same kind of absolute necessity as with just steps, but the situation is heavily weighted against large steps. This is so because *many* factors must come together to favor the persistence of any variety of creature, generation after generation; so even the smallest disturbance in some capacity to relate effectively to its environment places the survival of the line of creatures at risk. Evidently, the risk increases with the size of the disturbance, and natural selection will, except in the rarest of cases, simply lead to the extinction of those creatures that are the carriers of large steps.

And so, when I refer to "extremely small steps," I mean steps so small between one generation and the next that only very careful "comparison" of successive generations by the environment would show that there had been any step at all.

This poses a very fundamental challenge to evolutionary explanation generally since, in trying to follow some line of development, there will be no grand events of change to which one can point. The best one will be able to do is *imagine* some small step which, if amplified by a subsequent process of mutation and natural selection, would, after many generations, produce a large, clearly recognizable step, as would be apparent to the "environment," or, in the case where we are the recognizing part of the environment, apparent to our *unaided* senses. Many of what must be such amplified expressions of small evolutionary steps become evident by comparing the behavior of creatures that preceded us with our own, and so I shall be making such comparisons quite frequently, as has already been done in the foregoing section on instinctive behavior.

You might know that the question of big steps or small steps in evolution tends to rage as a great controversy from time to time. But the truly interesting aspect of this controversy is that it should occur at all, since, as I shall show in later chapters, it can be seen as a confirmation of the nature of brain, and the way in which it must have evolved. As we shall see, the really fundamental question turns out not to be whether there are big or small steps, but whether there are any steps, notwithstanding the apparent necessity of "extremely small" steps referred to earlier, based on there being discrete chemical elements. Indeed, we will see that this question derives from the same evolutionary feature of brain that gives rise to the controversy surrounding whether the universe "began" with some event such as a "big bang," that is, "began" at all, or is a manifestation of a continuous process having neither beginning nor end. So we can set aside the big-step small-step question for now, as being much more a manifestation of the quite local and, in evolutionary terms, recent nature of brain itself, than of the broader environment in which it must reside.

THE FUTURE AND GOAL-DIRECTEDNESS

At least in simpler creatures, a limitation to self-interest in an evolutionary world implies a limitation to the here-and-now. This implies an absence of any sense either of "the future" or of "goal-directedness" in them.

However, if an evolutionary theory is to get safely from simpler creatures to humans as we think we know them, the theory will, somehow or other, have to transition consistently from no goal-directedness to what we think of as human goal-directedness; somewhat paradoxically, one might reasonably insist that the appearance of a "sense of future" in creatures should arise from continuously applying the rule that the evolutionary process nowhere depend on a sense of the future! That is, a complete evolutionary theory should connect *itself* to what we experience as human behavior without needing to have the rules changed some very large part of the way down the road. How this might be accomplished is one of the major concerns of later chapters.

Absence of Precursors

If an animal has a normal life-span of five years and needs six months from birth before it can have young, we can expect to have about ten generations living at any one time. However, given the small steps in which evolution must proceed, it might have required thousands of generations for the animal to evolve from an identifiably different precursor to what it now is. This suggests that we can expect such identifiably different precursors to be *absent* from the population of animals that we see around us. Thus, when we say, for instance, that birds evolved from reptiles, this should not be taken to imply that any living reptile is, or even looks like, the reptilian precursor of birds.

Order in Growing

We are accustomed to seeing, in the usual growth of an infant, what is taken to be its "normal" course of development. But although it isn't so easy to see, there is also a normal course of development in a human embryo. Indeed, generally, there is a normal course of development for every kind of creature—just a single normal way of growing, particular to its kind. What, then, determines each of these particular ways of growing? Each of these particular ways of growing is determined by the *evolutionary history* of the creature. And it could hardly be any different, since evolutionary change can only occur in what already exists as previous evolutionary outcome; so, as a creature has come to being what it is, there is only one particular path along which it must have evolved—it can have had only a single, unique evolutionary history. And, since genes determine what normal growing will be, it is in genes that the unique evolutionary history of the creature is stored, and out of which comes the expression of this unique history. Looked at more mechanically, there is only *one way* that the creature can come together so that all the parts can be *sure* to fit—the way its predecessors went together when all the parts actually *did* fit. In evolution, all parts came after others, with natural selection filtering the order, mutation after mutation. And this has imposed on growing that some things *must* come after others, in an order that can change only at the risk of all the parts not *quite* going together, with natural selection vetting the ensuing debacle.

EVOLUTIONARY STAGES

Evolution seems to have addressed the two basic challenges to the survival of creatures in a sequence of overlapping stages. The first stage addressed parts and the nature of the parts—long and strong chemical substances such as proteins, membranes, shells, bones and so on. The next stage builds on this and addresses the working of the parts—automatic action of limited scope in various parts and groups of parts in the creature. The third stage addresses automatic activities in the creature as a whole—instinctive behavior of various types.

This is where, so far as I can see, the evolutionary story tends to come up against one of its most serious challenges, since, even if complex, instinctive behavior is marvelous to behold, as in insects, for example, this is a far cry from the kind of behavior we find in mammals, not to mention apes among them, and humans, in some sense or other, among these. If the evolutionary story is to save itself in being able to follow through to all of observable life, then another stage that goes far beyond programmed instincts needs to make its appearance. I believe that this next stage was achieved as a result of the appearance, within the evolutionary stream, of what we generally call "play."

CHAPTER II
THE EVOLUTIONARY ORIGIN
OF PLAY

The Paradox of Animals That Feed On Animals

One of the specializations to be found in the third stage of the sequence just mentioned is the group of animals that, in order to survive, consume the *same* types of complex animal materials as those of which they *themselves* are made. This specialization followed preceding living forms which could live on simpler materials: plants that had the capacity to combine the light from the sun with the minerals around them to build their own parts, and animals that feed on plants. These were followed by the widespread groups of animals that feed on other animals in order to obtain the materials for their parts; they include some fish, amphibians, reptiles, and birds as well as animal-eating mammals.

Before looking at the evolutionary paradox presented by animals that feed on animals, it is useful to look briefly at what this type of animal, in its most advanced state, must accomplish; and what this implies. Evidently, such an animal must be able to arrest and kill its prey before consuming it. This entails actions such as chasing, seizing, striking, clawing, and biting. If the animal is to be effective, it must evolve a suitable set of instinctive programs for controlling itself, including one which connects hunger to seeking prey.

And now we come on the basic paradox: If these animals are programmed in such a way as to link hunger with killing other animals, how could they have achieved and sustained the necessary reproductive capacity to avoid becoming extinct? What could have prevented them from killing their own particularly vulnerable young on a scale that would have so reduced the numbers of their offspring as to render them extinct, and so deny their very existence?

RESOLVING THE PARADOX

There are a number of ways out of the paradox. One way is for the animal to lay eggs which are left to hatch in the absence of parents, as with some fish and reptiles. The mobile, autonomous young are then exposed to their parents later, if ever, simply as another predator. Another way would be for the parents to die during the time between egg-laying and hatching.

However, among animal-eating animals, there are parents that raise their helpless young in close proximity to themselves, and so there must be yet another way of resolving the paradox. This must involve muting the killing instinct of these parents in the presence of their young. However, I should emphasize that such muting *cannot* be assumed to evolve so as to protect the young, since this is precisely the kind of future-based assertion which must be excluded from evolutionary explanation. How such muting comes to be in the interest of the *parents themselves* I leave till later.

It will be useful to summarize the argument to this point and then extend it a little. We can observe around us a wide range of animals that kill and eat other animals because they need the materials of which their prey are made in order to make possible the growth and maintenance of their own parts. The mere existence of such animals constitutes a paradox since, on the face of it, they should consume their young soon after birth when they are most vulnerable; and so these animals should simply not exist. By contrast, no such paradox exists for animals that consume only *plants* to meet their physiological needs.

In view of this surviving-animal-eater paradox, we can say that, somehow, for whatever self-serving reason not directed at preservation of young as such, some, at least, of these animal-eating animals must have evolved a procedure which caused their killing instincts to be muted in the presence of their young. This would certainly have to be the case for those groups of animal-eaters in which parents raise their feeble young in close proximity to themselves, and it is within this restricted group of young-rearing animal-eaters that, in my view, the origin of play is to be found.

THE ORIGIN OF PLAY

So picture now the scene in which such a parent is sitting among its young. Hunger drives the parent to do what it lives by doing: kill and eat. The specific characteristics of the young which identify them as related young now trigger a set of transformations in the parent which limit the force of pawing, limit the force of biting, limit the force of seizing and clutching. The young animals will, as the fighting killers they are in the process of becoming, defend themselves in their junior way. This continues until the parent moves away to find some other way to try to feed.

Watching the scene we would say how amusing it was to see the parent "playing" with its young; and in this way, I believe, we would have come on the essential nature of play. Indeed, for reasons that I shall explain later, I believe we would have come on what should be called "primal play."

With the foregoing as background I propose the following definition:

> Primal play is an activity that one can observe taking place between an animal-eating, young-rearing parent and its young. It is evidence of an attempt by the parent to kill and eat its young, but under conditions of internally generated attenuation of the parent's killing instincts. The attenuation is brought on by the presence of the parent's young, and is adjusted to such a level as just not to cause them harm.

This definition should not be taken to imply that the parent is, in any primary sense, acting to save its young. What might be driving the parent in a primary, self-serving way I leave till later.

Based on the definition, it would be expected that primal-play activity might include stalking, striking, seizing, clawing, biting, lifting and letting fall, shaking, beating, submerging, screaming, advancing and retreating; because each of these actions has its place in killing, and they all admit of muting in order to transform what would be almost certain death into what we see as play. However, if we interpret the definition at all carefully, we will see that the list just given of what primal play might include could be highly misleading, since it could only relate to a kind of animal that eats fairly large animals, rather than insects, say.

Thus, primal play will have to be taken to look like what killing would look like in the animal in question. So if the animal eats small lizards, its play can be expected to look like an animal killing a small lizard, and not like a lion trying to kill a zebra. In a general way, then, it is important not to take what play must look like as being necessarily similar to the play of dogs or cats or humans. To the extent that an animal plays at all, it plays in its own way. What the definition tells us is that this is as a muted version of how it would attack and kill the animals it eats.

There are still two major gaps remaining in this part of the speculation. The first has to do with the lack of any treatment of the tendency for animals to indulge in what might be called "exercise-play," as distinct from what I am calling primal play; the second has to do with the avoidance of any discussion of the self-serving reason for the muting of parental killing instincts, mentioned a number of times previously, and of which the survival of young is, at most, a side effect. I shall now address these in turn.

EXERCISE-PLAY

We can begin to get an understanding of exercise-play by noting that, to the extent that the available food supply will permit, excess physical activity, above that necessary for maintaining minimum physical and physiological integrity, will contribute to increased capacity for the animal to maintain both of these in itself. Thus, tendencies for animals to include in their behavior apparently pointless exercise activity—"unnecessary" physical activity in almost any form not associated with excessive risk—are entirely within the bounds of reasonable evolutionary development. Such activity might take the form of solo and group running, rolling, butting into inanimate objects such as hedges, jumping and twisting. Indeed, physical activity, in almost any form not associated with excessive risk, forms a basis for exercise-play. What distinguishes this from primal play is the lack of a strong interaction between parent and young, as well as the absence of those modes of behavior which can be identified with the parent's "real-life" acts of killing.

It is worth noting, however, that, because the basis for the advantage of exercise to the animal is so deeply buried in the complexities of its biology— witness how long it has taken us to demonstrate and accept this in ourselves— it is easy for such excess physical activity to have seemed pointless, even to careful human observers of play, and been confused with primal play; since, to some extent, primal play itself has the appearance of pointless, unaccountable activity.

A useful way of viewing all this exercise activity is as an evolutionary development which provides a way of employing excess food, when available, to increase the capacity of any animal to survive. Thus, we might reasonably expect to find displays of exercise-play in an unlimited range of animals, as opposed to primal play, which should appear in the behavior of animal-eating, young-rearing animals.

PRIMAL PLAY AS SELF-SERVING ACTIVITY

We can get at primal play as self-serving activity in the following way: Imagine a reptile which, as some reptiles do, lays its eggs and leaves them to hatch in isolation. Assume, also, that it feeds on insects, but that these, because of a change in climate, begin to be scarcer and scarcer. As the supply continues to decrease, insects, as sources of food that need to be sought and caught with speed and energy, become increasingly precarious, by comparison with the creature's own eggs.

It is therefore a development involving the self-preservation of a parent if its eggs—the easy "prey," that can be taken at will, even by an exhausted, starving parent lacking speed and energy—should, by a series of evolutionary shifts, come to remain as available food until the very last stage of hunger. It is to the *parent's* own primary advantage, and only *coincidentally* to the advantage of young, that some evolutionary tendency should develop in such a way as to attenuate the parent's own animal-eating instincts in the presence of its eggs, and only release them in the most extreme case, that is, when the parent is experiencing only the most extreme hunger.

Of course, as the evolutionary world unfolds, some parents will consume their eggs just a little sooner than others around them. Subsequent generations of these will simply fade away, as those with slightly more surviving young slowly capture the available resources and drive them, finally, to extinction. All that natural selection will lead to, eventually, is an array of insect-eating creatures that tend their eggs harmlessly…almost all the time.

It is not difficult to imagine that these creatures can evolve so that they no longer tend just their "supply" of eggs until they hatch, but begin to tend nests of their young. However, even when this development occurs, the situation is similar to what it was with eggs, since the young creatures also constitute a source of food which the parents can consume. The great difference that we can observe in external behavior comes about because, whereas eggs are totally defenseless, even young animals are not, and so the preservation of the parents' food-supply of last resort now takes the form of appearing to spare the young in an attenuated form of killing, in which the young can be seen to carry out their junior part in "play."

But it is important to understand that the play which we observe today must be quite different from that which was evolving during tens of thousands of generations in the precursors of the animal-eating, young-rearing animals that we now find us. What we can observe lacks all the intermediate steps, and especially those that terminated in extinctions. Further, even the animals that we have around us do not have their precursors among them, and so, since we cannot tell much about play from the record available in fossils, we are left to imagine a process that could have led to what we can actually see.

It is filling this gap, between what is, and what must have been, that is the challenge. And it is that gap, which, until we fill it with some imagining of a *consistent* evolutionary past, allows the buried workings of this past to challenge us with a paradox at the surface: a young-rearing, animal-eating animal that exists at all! Fortunately, we are not entirely without reference points, since there exist, among the animals known as "monotremes," creatures that resemble those in the foregoing sketch, in that they are fur-bearing insectivores that lay eggs, and suckle their young. Among these are the "echidna," about which I shall have more to say in Chapter IV, on *sleep.*

You have, I expect, begun to see the significance of this long evolutionary sequence for the way in which we should view the "gentle maternal scenes" to be found around us today: the household cat nursing its kittens, or, even closer to home, the analogous scene among ourselves. So since I know how difficult it is for us to entertain challenges to our belief in ourselves as fundamentally gentle and virtuous, especially as this relates to the nurture of children, it is useful to be reminded that, "paradoxically," as soon as human societies begin to produce more food than they need, they begin to produce less children.

This is just what the foregoing evolutionary development would predict, and, although we might prefer other explanations, which would portray us in a more kindly light, it is not easy to set one aside that leads us to resolve such a refractory paradox that marks the present, and which stretches over such a span of evolutionary past. This, clearly, is not to say that every nursing mother of today is waiting for the first opportunity to devour her young, but it does say something about the way in which a consistent evolutionary account would have us explain the "self-sacrifice" of suckling in mammals, and of the feeding of young by regurgitation in birds.

PREDICTING WHICH ANIMALS PLAY

As I indicated in Chapter I, explanation, to be really interesting and useful, must lead, at some point, to conclusions which admit of confirmation or denial—with emphasis on denial. What I plan to do now is see how well the foregoing explanations and conclusions stand up to this kind of "falsifiability" testing. The way in which I shall proceed is to predict which animals should play and then compare these predictions with the data gathered by careful observation of animals and their play as reported in Robert Fagen's excellent book *Animal Play Behavior.*

Before starting this comparison, I should explain that it has not been the practice, even among careful observers, to distinguish between exercise-play and primal play as I have just done; both types of activity have simply been treated together and referred to as "play," as will be apparent in the many quotations from Fagen. However, for reasons that might already be clear, but which will become even clearer in later chapters, this is an important distinction, and I shall be making frequent use of it in what follows.

I think it would also be useful to alert those who are not familiar with the observations and records of play to the fact that these are not like the observations and records of physics, or budgets and expenses around the house. In the case of play, what is being observed and recorded is much more complex, more vague, more resistant to capture in nets of numbers. And so, when we come to the comparing of observations with predictions, we should not expect the kind of clean precision with which we might be familiar in some other fields. This is not helped by the lack of resolution of play into even the two of its components to which I referred earlier. But there is not much room for serious criticism here, since what is really noteworthy in all these observations and records of play, which reveal so clearly the patience and care of life-long observers, is not their differences from those in other more quantifiable fields, but the fact that they exist at all!

We can begin prediction by assuming that only animal-eating, young-rearing animals will play. I shall provide a more elaborate version of this assumption a little later on, but this will suffice for now. So, among familiar animals, we expect that dogs and cats will play. But now we also expect hawks and owls and animal-eating birds generally to play. We would, by way of contrast, not expect that animal-eating turtles which leave their eggs to hatch remotely in isolation would play, because the young will simply disperse and resolve the paradox this way.

An interesting case is that of those (viviparous) lizards and snakes which bear live young rather than lay eggs. The young are generally miniature copies of their parents, and, being active soon after birth, disperse. Some of them might be eaten, but most would escape by chance. The paradox is therefore resolved in this way, and such animals would not be expected to play. In this way, we come on our first major prediction:

Reptiles will not play.

I can think of no fish that would play, because they either lay eggs which hatch remotely, and so resolve the paradox this way, or they have young that are born alive, again as active miniatures of their parents, and so have an opportunity to spread out soon after birth, escape being eaten, and thus resolve the paradox in another way. This leads to the second major prediction:

Fish will not play.

It is encouraging when a prediction leads to what one might think to be likely; but it is much more encouraging when it leads to something that one would not otherwise think to be so, as is the case with the predictions that fish and reptiles should not play. To see this, one need only read the following quote (Fagen, page 219):

> "Despite the highly sophisticated forms of social (including parental) and feeding behavior found in many species of

fishes and social insects, play is virtually nonexistent in these small-brained and relatively decorticate species . . . Even in reptiles, where play might confidently be expected purely on phylogenetic grounds, especially in the large, long-lived varamid lizards or in family-living crocodilians, accounts of play are rare and restricted to interactions with objects..."

It is useful to look carefully at the last sentence: "accounts of play are rare and restricted to interactions with objects . . ." Using the distinction between types of play which I have been making, the sentence can be rewritten as follows: "there are no accounts of primal play, and the accounts of exercise-play are rare and restricted to interactions with objects . . ." This is perfectly in line with the prediction, since it is primal play that is severely limited in its occurrence and should not be present in either fish or reptiles, and exercise-play that we can expect to find universally. Thus, the apparently perplexing observations recorded by Fagen regarding fish and reptiles are not so perplexing at all, but, instead, flow naturally out of the present speculation.

Since insects are mentioned in the quotation, I shall include them here in the prediction by saying that the presence or absence of play will depend primarily on whether any particular insect does or does not eat animals. If the insect eats animals, it will or will not play depending on whether the young can or cannot escape the predatory assaults of parents by crawling away or otherwise escaping, as with fish, for instance.

Notice that there is nothing about either reptiles or fish generally which suggests that they will not play. To arrive at the conclusion that reptiles, for instance, will not play, the argument has to proceed group by group within this large class. We begin by excluding those that eat other animals but lay eggs which hatch remotely, and then those that eat other animals and bear live young, since these are sufficiently autonomous at birth to escape the predatory assaults of parents. We can then set aside those that eat plants, since, if the animal-eaters among them will not play, neither will the plant-eaters. Having exhausted the possibilities, we conclude that reptiles, "as a class," will not play. The conclusion that reptiles will not play is of very far-reaching significance in much of what follows, and so it is important to understand the definitive way in which this conclusion is arrived at.

If we move now to plant-eating (herbivorous) mammals, the prediction would seem to be that such animals would indulge in exercise-play only. But it is here that the elaboration of the assumptions referred to previously needs to be made, and the reason for this is as follows. It is well known that the post-reptilian creature that was the ancestor of mammals was an insect-eater which reared its young, in a way that, presumably, resembled the monotremes of today. Such a creature must, according to the basis of play being advanced here, have indulged in primal play. Now, as I shall show in following chapters, the emergence of primal play in a creature endows it with certain other behavioral possibilities which favor the evolutionary retention of play behavior, *even when* the animal-eating basis for its emergence might have disappeared. And since primal play is *certainly hereditary* at this point of the mammal's evolutionary development, some version of such behavior might continue to be transmitted from one generation to the next, *even in some of those mammals that subsequently become plant-eaters,* and thus become separated from the basis for the emergence of primal play. Hence, we should find, among mammals, at least some plant-eaters which display behavior that could be interpreted as "vestigial" play.

The conditions for primal play in mammals then become that all animal-eaters must play, some plant-eaters might play, and play can be totally absent only in plant-eaters. This last condition reflects both the total disappearance of even vestigial play derived from the insect-eating young-rearing mammal ancestor, as well as the lack of any current need to seek food by killing. Since the reptilian derivative that was the ancestor of birds was an animal-eater, we can expect the same conditions to apply to primal play in birds as in mammals.

We can now return to comparing observations of play in mammals with predictions, and here is what is observed for plant-eating wild cattle (Fagen, pages 196-7):

> "play in the wild cattle (Bostaurus) of France's Camargue region includes solo frolics by young calves, who run around, leap vertically and twist their head or body while in midair and kick up their heels (Scholoeth 1961). They may butt or try to mount their mothers (Scholoeth 1958). Older calves and yearlings in the Camargue herd played together around a small hillock. They would butt, rear, and 'box' with their front legs and chase and flee. Playfights were decorated by springs and capers, and boxing and butting alternated with chasing and fleeing. Calves also horned bushes and hummocks (Scholeth 1961)."

As predicted, there is clear evidence here of exercise-play, but there is relatively little that corresponds to primal play which would include muted attempts at killing. This has no place in the lives of these plant-eating mammals, and so even vestigial play seems to have almost completely disappeared from their behavioral routines. All of this is in good agreement with what one would expect.

Frequently, it is at the transition between extremes that real interest lies, and the domain of play is no exception. So let us turn now to that wide transition class of animals which eat both plants and animals. We can predict that play activity should increase as we go from the animals that eat mainly plants to those that eat mainly animals. This can be observed as shown by the following quotation (Fagen, pages 81-2), in which the animals go from plant- to animal-eaters as it goes from start to finish:

> "The family Dasyuridae has radiated widely. It includes species resembling mice or small rats, as well as carnivorous species. Among these carnivorous marsupials are the viverrid-like native "cats" (*Dasyurus* Epp.) and the small, bearlike Tasmanian devil (*Sarcophilus harrisii*). Behavior of several capture-bred dasyurid species have been studied. Play of the young in small (head and body length of adults ca 100 mm), mouselike dasyurids (*Sminthopsis murina* and perhaps *Sminthopsis crassicaudrata*) is restricted to jerky running and brief dashes, and social play is absent (Ewer 1968b, Graham Settle pers. comm...). Young of the larger (120 to 250 mm), ratlike forms (*Antichinus stuartii, Dasyuroides byrnei*) exhibit "erratic leaping and bounding" (Aslin 1974) and sporadic, simple social play which principally involves chasing and which frequently escalates into agonistic fighting (Settle pers. comm..). Young of the larger (300 to 500 mm) viverrid-like dasyurids (*Dasyurus hallucatus, Dasyurus viverrinus*) play by themselves, chase and wrestle together, and play with objects (Fleay 1935, Nelson & Smith 1971, Settle pers. comm..). In *D. Viverrinus*, the larger of the two species, mother and young continue playing with each other after weaning (Nelson & Smith 1971), and tame individuals may remain playful all their lives (Fleay 1935). *Dasyurus maculatus*, the tiger quoll, the largest (600 mm) and heaviest (6 kg) of the native "cats," is a mustelid-like stalking predator inhabiting heavily timbered areas and rain forests. Chasing, wrestling, and stalking social play ... is frequent, is structurally complex, and rarely escalates in these animals (Settle 1977, pers. comm.)."

I should emphasize that the Dasyuridae are marsupials, that is, they belong to that large order of animals in which the early phases of development of a new-born take place in a pouch (marsupium) located outside the abdomen of the mother. The marsupials run almost parallel to the more numerous placental mammals, and so confirmation here adds an extensive group to those included in satisfactory prediction.

Birds are a large and interesting class, and so one wonders what they might do by way of play. The first part of the prediction says simply that all the birds that eat insects, and mice, and fish, and other birds, and rabbits, or whatever other kind of animal, will play. They all will play because the eggs of birds are incubated by the birds themselves, and so the emerging young, which are often totally helpless, are exposed to being eaten by their parents.

In this respect, animal-eating birds are completely different from animal-eating reptiles which leave their eggs to hatch remotely or, in the case of those that bear live young, have newborn that are operating miniatures of themselves which can escape in sufficient quantities to resolve the paradox without the occurrence of primal play.

The evidence for play in birds is suggestive but unclear as indicated in the following quote (Fagen, page 218):

> "... most avian play seems somewhat ambiguous by mammalian standards, either because it has not fully emerged from maturational intermediate status or because existing accounts relied on inadequate criteria and failed to provide necessary information."

I read these observations as calling attention to two related and important cautions to which I have already referred. The first has to do with using "mammalian standards" for assessing play in birds (or any other non-mammals), and the second with the related likelihood of having "relied on inadequate criteria." I believe that these cautions apply fully, and that the present speculation allows one to address them both, since it provides a basis for developing adequate criteria which would, evidently, flow from noting what killing behavior looks like in various kinds of birds, and how this might be expressed when muted. These, coupled with recognizing the distinction between primal play and exercise-play should go a long way towards what is required.

It is therefore easy to agree with the following statement (Fagen page 218):

> "Avian play provides a comparative base-line essential for understanding the evolution of animal play behavior. Birds

> could become as important to the study of play as they
> have been to developmental studies of such phenomena as
> imprinting and species-specific vocalization."

I believe that when, in time, the required observation is done in a sharper theoretical context, it will be found that birds such as hawks, eagles, owls and similar clear-cut animal-eaters play in unmistakable fashion. At the other end, plant-eating birds, such as geese, as well as smaller plant-eaters will be found to play only little, if at all. Between these should be a whole intermediate range of play.

Another family of animals, in which play behavior should provide a basis for comparison with the prediction, is the primates, which includes tarsiers, lemurs, monkeys, apes and ourselves. The general prediction would be that the members of the family that eat plants alone might not play, and those that eat plants and animals would. The ones that eat the largest animals would tend to show the clearest signs of play.

The dietary information for a number of the primates is as follows:

gorilla,	fruit and shoots;
orangutan,	fruit;
gibbon,	leaves and fruit;
rhesus macaque,	fruit, insects, lizards;
chimpanzee,	fruit, eggs, insects, large animals;
baboon,	fruit, eggs, insects, lizards, large animals;
human,	fruit, insects, reptiles, mammals, and much else.

The detailed prediction would be that the gorilla, orangutan and gibbon would exhibit behavior somewhere between vestigial play and no play, while the chimpanzee, baboon and macaque must clearly exhibit play. The human should also play. We can begin with the following quote (Fagen, pages 102-3):

> "The great apes are among the longest lived and most
> highly encephalized primates, and by all rights they should,

without exception, exhibit play. This is certainly the case for healthy, but confined animals in naturalistic environments, but paradoxically field workers have remarked on the low frequency of play in populations of at least three great apes: gorillas (Gorilla gorilla) (Schadler 1963), orangutans (Pongo pygmaeus) (Horr 1977), and, despite Carpenter's extensive description (1940) of their play in the wild, gibbons (Hylobates lar) (Ellefson 1968). The message of this literature is that the young of these three great apes are solemn, incurious, and unresponsive creatures."

On this basis, the first three predictions would appear to be confirmed, even "paradoxically" so. But the section quoted continues as follows:

"… As anyone familiar with young orangutans, gorillas or gibbons can testify, nothing could be farther from the truth. Consider, for example, the play of young orangutans in a family group (Mackinnon 1974a, Maple and Zucker 1978, Wallace 1962, Zucker, Mitchell, and Maple 1978, pers. obs.). These animals wrestle on the ground, chase, bang and wrestle, drape themselves with various inanimate objects, and exhibit playfaces. In confinement, an indulgent father may push his daughter over with his hand, roll her back and forth on the ground, and even tickle her. Siblings of very different ages play together; the elder may take a passive role or may use less than its full strength in play."

As you will doubtless have noticed, I find myself stranded between experts. At least some of the difficulty comes from the lack of distinction, as discussed earlier, between different types of play-behavior. Other difficulties almost certainly arise from comparing what "field workers" observe, as in the first part of the quotation, with what is observed with animals "in confinement," as in the second part. However, I believe that the apparent confusion is a reflection of more complex factors than these, of which by far the most basic have to do with vestigial play, and extensions of it which are discussed fully in Chapter IV, on *sleep*.

So I shall take the observations attributed to "field workers" in the first part of the quotation, since these relate to the environment that is relevant to the present speculations. When this is done, the observations match the prediction, a condition which, *by itself*, is not of much consequence, but which, as part of an accumulating number of matches of this sort, clearly is.

The prediction regarding the chimpanzee, baboon and macaque is that they must play, and to much the same extent. The following quote (Fagen, page 102) is useful as a start:

> "Baboon play does not differ greatly from rhesus play (Altmann & Altman 1970, Owens 1975 a,b, 1976, Remson & Rowell 1972, Rose 1977 b)."

The following two quotes are also relevant:

> "… Play of juvenile chimpanzees with conspecifics, with baboons, and with manipulable items, as well as chimpanzee solo and mother-infant play, have been described lovingly and at length by Jane Goodall (VanHowick – Goodall 1965, 1967, 1968 a, 1970, 1971, 1973) and by other observers (Clark 1977, Loizos 1969, McGrew 1977, Menzel 1963, 1975, Savage & Malick 1977, Simpson 1978, Jutin and McGrew 1973)," (Fagen, page 106);

and:

> "Mother chimpanzees play with their infants, tickling them and nibbling at them." (Fagen, page 109)

As we all know, the human plays, and this completes the comparison between the predictions and observations for primates.

You might have noticed the references to "tickling" in some of the latter quotes, and so it is interesting to seek what the evolutionary connection might be between tickling and play. Since we associate tickling with *laughter*, and I am associating play with the threat of killing, it would seem difficult to see what this connection might be. But this is because we tend to associate laughter with *amusement*. However, as the discussion of sleep in Chapter IV will show, there are good reasons for believing that laughter is an evolutionary concomitant of *stress*, rather than of *amusement*.

If this is so, we can see what the connection between tickling and play must be, since it would mean that the origin of tickling should be associated with some stress-generating threat. That is, of course, exactly what we would expect if tickling were an attenuated form of killing behavior, as well it could be, since it resembles the killing behavior of an animal that kills insects and small reptiles like lizards, for instance, as such behavior would be applied to a larger animal in play.

What all this would mean, then, is that tickling forms a very natural part of the repertoire of attenuated killing behavior that we see in play, and that the associated laughter is simply a part of the general indication of stress experienced by a young creature exposed to the threat presented by play.

It seems to me that, when one includes an explanation such as that just given for tickling, the evolutionary origin of play given here is fully matched by the observations of play that have been recorded by numerous devoted and careful observers. Indeed, the match is satisfactory over an enormous range of classes and species of creatures, and will be seen to be even more so when developments in later chapters can be used to clear away the few remaining ambiguities.

In summary, it seems reasonable to assert at this point that the evolutionary origin of play is to be found in the emergence of animals that both eat animals, and rear their young in close proximity to themselves. Thus, by far the clearest evidence of play to be found today is in such animals. However, since such play would have been present in the earliest ancestors of mammals and birds, because they were animal eaters, and reared their young in close proximity to themselves, and since it would have been hereditary, traces of play might still be found even among mammals (and, I believe, birds) that now eat plants alone. Traces of play in such plant-eaters, when it can be identified, is a hereditary vestige of the primal play in which their ancestors indulged.

Whether we are observing "true" or vestigial play, what we see in play is the attenuated form of killing behavior in which either the playing animal itself still indulges, or in which some ancestor indulged. It is useful to distinguish this kind of play, which I have called "primal play," from another kind of activity, which I have called "exercise-play," and which is to be found in all animals, since it represents simply a way of storing surplus food as survival potential. As a result of the evolutionary origin of primal play, it will not be found among either fish or reptiles, but will be found among mammals (and birds), mainly according to the details of their diets.

Although such an evolutionary origin of play is interesting in itself, and could be pursued in greater detail, there are many other interesting aspects of play to which we should turn, as the following chapter shows.

CHAPTER III
LEARNING TO COPY IS PLAY

BEYOND PRIMAL PLAY

One of the important conclusions arrived at in the previous chapter is that primal play marked an identifiable divergence in behavior between reptiles and the animals that evolved from them. In this chapter I am going to show how the behavior that forms a normal part of primal play comes to include just those very critical components that allowed mutations and natural selection to lead ultimately to clear instances of behavior based not on heredity and instinct alone, but on *copying* as well.

COPY-LEARNING

There are some creatures, turtles for instance, that are hatched from eggs in total isolation from parents or siblings, but which develop the full range of behaviors necessary to cope with their environments in the ways that are normal for members of their species. Clearly, the behavior acquired in this way must be determined by genes that are parts of the eggs from which such creatures develop. However, in some other species, of which ours is the best example, we can observe, in their members, behavior that cannot be acquired except in the presence of at least one other member *that displays the behavior to be acquired.* An example of this is the language spoken: Chinese, French or whatever. So equally clearly, the behavior in this second case *cannot* be hereditary, since it need not resemble behavior in a parent or any other ancestor, and will not be passed on to offspring without the same kind of *social* contact that started its spread. What we can say about this kind of behavior is that, although the behavior *itself* is not determined by genes, *the ability to acquire behavior in this way must be determined by genes.* I shall call this the ability to "copy-learn," and what is acquired in this way "copy-learning."

Since the first kind of behavior is determined by genes, and the *ability* to copy-learn is also determined by genes, we have two *different* kinds of behavior determined by genes. These must, therefore, be different genes. But different genes can only be present if some evolutionary process led from one kind of gene to the next, and so the question arises as to where, in the evolutionary stream, creatures ceased having just genes that control hereditary behavior, that is, ceased having instinctive behavior only, and began to have, *in addition,* genes that made copy-learning possible. It is easy to see that this must have been the actual order of evolutionary development, since no creature could survive that did not have some minimum hereditary behavior to tide it over the first few minutes of life, at least.

As indicated above, it appears that primal play was exactly the new element of the environment required to induce the emergence of copy-learning, and that is what I shall now set out to show.

OFFENCE AND DEFENSE IN PLAY

In what follows, there will be frequent references to "primal play" and to a "playing animal." To avoid writing them out each time I shall use "play" for the first and "playal" for the second.

So imagine a playal parent that attacks its young and initiates play. This evidently provides the young playal with an opportunity for acquiring new behavior. What it will acquire most directly is behavior for defending itself, but it is the *conditions under which it will do so, that are unique.* The conditions are unique because the play attack is *a muted version of the parent's* "real" *behavior.* This being so, the young playal can repeat, what is for *it*, a real-life form of behavior at virtually no risk, since the consequences of failure to defend itself adequately are limited to a chance to try again! This is what I have referred to previously as "the unique evolutionary shade of play."

In all the generations that had preceded the emergence of the playal, failure of the young in acquiring the behavior of defense would lead to injury and even death. But now, in the remarkable new environment of the young playal, there exists a meaningful but strangely *artificial* version of *real* components in its life. Indeed, one can view the world of play, somewhat paradoxically, as the *first,* **real***, world of the* **artificial**—*the first real worlds of "make-believe" and "as if"*—and the young playal as the first animal ever to find itself in such a world. As just mentioned, this certainly improves the conditions for acquiring the behavior needed for defense, but that is not what play provides that is *quali*tatively new. To see what this might be we need to look at how the playal can improve its *off*ence, rather than it *def*ense.

To see this, notice that what the young playal must acquire for *offence* is behavior that allows it to do exactly what its parent is doing *to* it in play; so what it needs to be able to do is *copy* the behavior of the parent, not simply oppose it in a series of repetitions, as it does beneficially in defense. A way in which it could do this kind of copying is to carry out the following operations:

1. Isolate the view of the parent attacking in play from everything else around, as far as possible.

2. Store an image-*sequence* of the isolated parent attacking in play.

3. Use the sequence of *behavior* of the parent in (2) to control the behavior of the young playal in play-attacks on its parent.

It will be convenient to place the discussion of these three operations within three larger sections as follows:

A. Which looks at the operations themselves, and how they might be realized as reasonable biological processes;

B. Which discusses a number of aspects of emerging copy-learning that have significance for other phenomena which only become fully established in later generations of creatures, such as ourselves;

C. Which summarizes the reasons, as they appear in A, for which the situation of the playal was uniquely suited to the emergence of copy-learning.

Section A: The Operations in Copy-Learning

Operation (1)

"Isolate the view of the attacking parent in play from everything else around, as completely as possible."

The reason for this operation is to be found mainly in the operations that follow, which, because of their complexity, require that the image of the parent be as free of extraneous material as the biology of the playal can render it. What is required of the playal we could call "attention," and the situation of play is ideal for providing this since, while the young playal is under attack in play, its instinctive program focuses the whole of its attention on the field occupied by its single attacker. Thus, the operation is performed, in a powerful and *automatic* way, *by the nature of play itself.*

Of course, other situations, in which the young playal could be under *real* attack by some random creature *other* than its parent, would also serve to focus attention on the field occupied by its attacker alone, but these are not situations in which the repetition of novel forms of behavior can be performed by a young playal at as low a risk as in play. Furthermore, it is only when the attacker is the parent of (the same species as) the young playal that what would be copied can, with any chance of a copy being realized, be exactly the behavior of its attacker, due simply to practical species-specific behavioral limitations in the playal. This is of great importance for such a complex series of operations which, especially in this *emerging* stage, can benefit from *every* simplification, such as that of not having to resolve which parts of a behavioral routine are to be copied, and which not.

Related to this kind of simplification is the additional fact that random attackers would have random images which pose what would seem to be an insuperable problem for the evolutionary *emergence* of a copying process, since what is to be copied would be changing from one attack to the next. This problem is almost totally absent in the case of play, in which the same parent (or, at worst, the same two which resemble each other) will launch the same attack and generate almost the same image from one day to the next.

Operation (2)

> "Store an image-sequence of the isolated parent attacking in play."

The word "image" provides a good starting point for discussing this operation, since it tends to suggest "visual" image. However, the "images" that are intended here are those made up of nervous streams originating not only in vision, but in the senses of touch, smell, taste and hearing as well. So these images are composed of nervous streams arising from the changing relation of the playal to the world outside it, including what it can see, and hear, and touch, and smell, of *itself*, as conveyed by its sensory activity. It is convenient to call this stream the "outer" stream.

This image-flow must be accompanied by a *second* stream arising *inside* the playal which continuously describes the state and relationship of those portions of itself, mainly muscles that are involved in its external behavior. It is easy to see that every creature that possesses a brain, as part of a central nervous system, must have such an internal way of informing the brain of the relation between the various parts of the creature involved in external behavior, since, without such information, the creature would ultimately be destroyed by the efforts of a brain directing muscles that were not actually in the state the information available to it indicated they were in. It is convenient to call the stream that provides this kind of internal configuration information the "inner" stream.

Now all of us are familiar with outer streams, or at least the "images" associated with them, but we might not all be so familiar with the fact that there exists a whole network of nerves devoted to producing the signals that go to make up inner streams. The nerves making up this network are often referred to as "proprioceptors," and form part of what is usually known as the "afferent" segment of the nervous system, where "afferent" is intended to convey the notion of "conducting inward" toward the brain. (The "efferent" segment of the nervous system conducts outward, away from the brain.)

An easy way of demonstrating the existence of proprioceptors and their function is to *close* your eyes and touch the tip of your nose with one of your fingers. As you will notice, this is possible *without* having to *see* your hand as it moves, that is, the relative positions of your hand and nose are given by an *internal* source of information which continually defines your (external) configuration *in your brain.* (If you have difficulty touching your nose in this way, you should have a doctor see why, since it could be a far from trivial symptom.) A beautifully clear description of proprioceptors is given in Chapter II of *The Senses of Man* by Joan Steen Wilentz.

So now I am going to assert that, although playals were certainly not the first creatures to have proprioceptor networks, they were the first (and only) creatures to store and use the information available in these networks as inner streams, in conjunction with outer streams, *in such a way as to lead to the emergence of copy-learning.*

I shall refrain from presenting the evidence in support of this assertion until a later point at which the development of the argument for copy-learning is further advanced. Here, I shall simply state, by way of summary, that an image-sequence stored by the playal of the parent attacking in play consists of two streams, of which the outer carries images of the parent and, of necessity, the *playal itself*, while the inner carries streams of nervous data which are records of the neuromuscular activity that correspond to the behavior of the playal as this is carried in its outer stream.

It is important to notice, for following portions of the discussion, that the playal's outer stream carries *no neuromuscular* information concerning the actual behavior of either the playal or its parent, and is limited to sensory information from which the behavior of neither the playal nor its parent could be reconstructed. By contrast, it seems reasonable to assume that, if some version of the playal's *inner* stream were coupled back into the (efferent) part of its nervous system which drives behavior, the playal would repeat the behavior which accompanied the inner stream originally, and hence *its own* part in its outer-stream images.

What this assumption implies is that the neuromuscular system of the playal is such that the nervous record of a certain muscular sequence, as produced by proprioceptors, would reproduce that same sequence if it were (re-formatted, and) re-introduced some time later as a signal for driving the muscles which had produced the nervous record in the first place. It is difficult to imagine how a central nervous system could function at all without such consistency and reciprocity between signals that record its muscular behavior and those that drive it.

Operation (3)

> "Use the sequence of behavior of the parent in (2) to control the behavior of the young playal in play attacks on its parent."

The distinction between inner and outer images allows us to see clearly what the *primary* difficulty in copy-learning happens to be, since, *on the face of it*, what this operation would require is that the behavior of some actor, in this case the parent, located in the playal's outer image, should *drive* its behavior, so that the behavior of the playal can "copy" that of the parent which would then be driving its behavior.

But such a *direct* connection, as might be suggested by this operation between the behavior of the parent, as an actor in an outer image of the playal, and the behavior of the playal itself, is *clearly impossible;* since, as I have mentioned before, there are no records of the streams of neuromuscular activity that occurred in the parent available to the playal which could, in any one-to-one way, tie the specific behavior of the parent in the outer image directly to the behavior of the playal.

The record of nervous streams corresponding to the external behavior of the parent resides in *its own inner* stream; and, similarly, the only record of actual nervous activity available to the playal is in *its* own inner stream, and this describes *its own* behavior exclusively in terms of its own internal nervous streams, and nothing else.

It is clearly impossible to link these two separate inner streams, resident in parent and playal, by any imaginable biological process, and so, for copy-learning to emerge, some indirect process must develop which allows the image of the parent in the playal's outer stream to appear *as if* it were part of the playal's inner stream, that is, *as if* it were identifiable directly with the nervous system of the playal itself.

The key to this indirect process resides in the fact that the outer streams of the playal, gathered in what is for it defensive play, *inevitably* contain fragmentary images of the playal *itself*, in *addition* to images of its parent in offensive play. The process of copying the offensive behavior of the parent can then be transformed into an imaginable biological process of *making two images match,* as I shall now explain.

The fragmentary *images* of the playal *itself* in its own *outer* stream are authentic records of portions of its *external* behavior which *correspond* to its *nervous inner stream*. Therefore, what the playal needs to do to accomplish copy-learning is come on some series of actions that bring the behavior of its image, in its own outer stream, into correspondence with the behavior of the image of its parent; since, if it can accomplish this, it would have a behavioral routine, as recorded in its own inner stream, which could then reproduce its own image-matching behavior and hence, indirectly, the offensive behavior of its parent. And in this way it would have *copied* the offensive behavior of the parent.

It is useful to notice that there are two identifiable processes involved in making images match in this way. The first is that in which, using the images of its parent and of itself in its outer stream, the playal compares the behavior of its parent with its own, so as to identify a difference; the second is that in which the playal uses the difference to guide small changes in its behavior, which are recorded in its inner stream, so as finally to reduce the differences between the two images to some amount which it can reduce no further. If the playal could realize these two processes, it would be in possession of a nervous sequence in its *own* inner stream which could be stored and used to drive its *own* behavior, and so copy that of its parent, as this relates to *off*ence.

You will probably be surprised, now, when I point out that all the playal needs do, to begin to realize both these processes, is execute a behavioral routine, which continues to look like a simple extension of its defensive behavior in play, in which it opposes the offensive movements of its parent with *blocking* movements of its own, as when paw opposes paw directly, and open mouth seizes open mouth. When this state is reached in play, the playal has realized *both* of the required processes, since its own visible, opposing movements constitute not only a way of bringing the images of parent and playal together for comparison, but also a way of ensuring that the behavior of the playal has made them match.

When this occurs, the playal has come on a way of producing its required inner stream, and hence of copy-learning the offensive behavior of its parent by means of the only kind of change that natural selection will sustain, that is, by an almost imperceptible extension in its defensive behavior.

It is one of the remarkable features of primal play that the playal, in *defending* itself, comes to execute, over and over again, the repeated *offensive* behavior of the same attacker, and so takes the first step in the direction of being able to store and reuse its inner streams which, in recording the behavior of the playal itself, also record the behavior necessary to copy-learn the behavior of some other actor, outside the playal.

Evidently, what the playal has begun to do is use its own image (in the most general sense of "image" mentioned above), as it occurs inevitably in its own outer streams, as a sort of flexible template which can serve to transfer, with what must be varying degrees of accuracy and completeness, the behavior of an external actor into itself. It is important to notice that the playal does this by matching *images*, for this will allow it later to copy-learn "behavior" without the need for a real external actor, but by comparing the image of itself with the image of an "actor" that exists nowhere else than *in its brain*.

The ultimate power of the process is matched only by the necessary simplicity of its evolutionary emergence, and so I call attention here again to the remarkably ideal conditions that play provides for the emergence of such a major qualitative shift in the capacities of the creature. Clearly, the acquisition of offensive behavioral routines contributes directly to the ability of a playal to secure food, and so we can see how the forces of natural selection would favor the development of what later becomes a complex process of copy-learning that so directly addresses survival needs.

But we should also notice that what appears, ultimately, as such almost unfathomably complex developments in copy-learning, could begin to emerge as only the very smallest shifts in behavior that occur under the simplest of ideal conditions. To see how ideal is play for such emergence it is useful to notice the simplification that comes to the *beginnings* of the process by virtue of the *physical* resemblance of the playal to its parent. Notice, in particular, that, because of this resemblance, for every behavioral sequence in the *parent* that the young playal can experience and store in an outer stream, there is a *corresponding inner* stream, which actually originates and is recorded in the playal, but which could be treated *as if* it had originated in the parent, with *increasing* accuracy, since both the physical and the behavioral features of the playal and its parent converge more and more closely, as the process of copying becomes more and more complete. This convergence, of the physical and behavioral features of the playal on those of its parent, renders the initial attempts at image-matching much simpler than they would otherwise be.

It is, in my view, nothing but the almost perfect and continuously improving identity between the playal and its parent, which occurs uniquely in play, that renders the very first step in this process of copy-learning *behavior* by the comparison and matching of *images* even definable, and which places it within the realm of imaginable biological execution. It is the unique context of play which makes the beginning step small enough for natural selection to treat it as anything but a passing dysfunction, leading to extinction.

You must, at this point, have noticed that we have here come on the very first trace of an evolutionary turning point of the kind that I referred to in Chapter I, which signals the beginning of a long series of far-reaching developments, but is marked by a shift in behavior that is so slight that it has required pointing to the minutest of behavioral detail in order to bring it to your attention. Because of the very small mutations to which enduring evolutionary changes are limited, the situation can never be any different—an evolutionary change, which survives the filter of natural selection, will always have a small beginning. And the only way to have any confidence in the significance of such a small beginning is to see *where the assertion of such a beginning leads in achieving further* explanation of *still later* evolutionary phenomena. As you will see, the small beginning I have proposed *does* lead to the explanation of whole groups of later phenomena, to some of which we can turn immediately.

Section B: Copy-Learning and Some Later Phenomena

COPY-LEARNING AND ABSTRACTION

The process of copying can be viewed as separating the concrete details of the parent from its behavior, which then becomes available as a pure behavioral program, thus allowing the playal to repeat the behavior, but without assuming the actual particularities and concrete details of the parent. Viewed in this way, we might say that the playal has produced an "abstract" version of the parent, having abstracted its behavior, as pure behavior, form all its other real, concrete features.

This ability to perform abstraction is evidently a fundamental feature of copy-learning, since, because of the role which behavior plays in the process of evolution, the ability to effect the abstraction of behavior in particular, as pure reproducible behavior, as the playal begins to do with its parent, is of fundamental importance for its subsequent development, leading ultimately to its significance for us.

COPY-LEARNING AND THE SPREAD OF CHANGE

It might seem strange that copy-learning should be associated with the spread of *change*, but this is very much the case as can be seen in the following way. Imagine a parent P with a child C and two parents GF and GM, who are the grandfather and grandmother of C. Then the behavior of P might display some aspect bP which is not present in the behavior of either GF or GM, and which is also not present in the behavior of C, so far as this would be determined by genetic factors, that is, by factors other than copy-learning. However, if C now copy-learns aspect bP of the behavior of P, not only can it begin to execute bP in its own behavior, but it can become a new behavioral focus from which bP can spread to other creatures that can effect copy-learning and that come into contact with C. The spread of bP can also extend to the children of C, and to the children of the creatures with which it comes in contact, and so on.

Evidently, as bP is spread by copy-learning, *errors* will occur in copying, and bP will become bbP, say, and this will be spread by copy-learning, becoming bbbP, say, and so on, until the copy-learned derivatives are no longer easily recognized as being related to bP. What should now be clear from this is that copy-learning sets in motion a sequence of evolutionary development of behavior which is analogous to that which depends on heredity and mutations in genes—genes support one kind of transmission of behavior by copying, and copy-learning supports another. However, copy-learning will generally achieve the spreading of changes much more rapidly.

What is also clear is that this kind of evolutionary development, which has led to behavior that is distinctly evident in us, began to appear in our very earliest playal ancestors some hundreds of thousands of generations ago. This kind of spread of change is frequently referred to as "cultural transmission," especially as it occurs in us, but it should be observable in other playals. This is indeed the case, as the following quotation from *The Selfish Gene* by Richard Dawkins shows clearly:

> "Cultural transmission is not unique to man. The best non-human example that I know has recently been described by P. F. Jenkins in the song of a bird called the saddleback which lives on islands off New Zealand. On the island where he worked there was a total repertoire of about nine distinct songs. Any given male sang only one or a few of these songs. The males could be classified into dialect groups. For example, one group of eight males with neighboring territories sang a particular song called the CC song. Other dialect groups sang different songs. Sometimes the members of a dialect group shared more than one distinct song. By comparing the songs of fathers and sons, Jenkins showed that song patterns were not inherited genetically. Each young male

was likely to adopt songs from his territorial neighbors by imitation, in an analogous way to human language. During most of the time Jenkins was there, there was a fixed number of songs on the island, a kind of 'song pool' from which each young male drew his own small repertoire. But occasionally Jenkins was privileged to witness the 'invention' of a new song, which occurred by a mistake in the imitation of an old one. He writes: 'New song forms have been shown to arise variously by change of pitch of a note, repetition of a note, the elision of notes and the combination of parts of other existing songs… The appearance of the new form was an abrupt event and the product was quite stable over a period of years. Further, in a number of cases the variant was transmitted accurately in its new form to younger recruits so that a recognizably coherent group of like singers developed.' Jenkins refers to the origins of new songs as 'cultural mutations'.

"Song in the saddleback truly evolves by non-genetic means. There are other examples of cultural evolution in birds and monkeys …"   (Dawkins, pages 203-4.)

It is interesting that Dawkins' "best" example should involve a bird, since, as I pointed out in the previous chapter, there is every reason to believe that playals are to be found among birds, even though there is no *direct* recorded evidence (yet) in support of this. The interest extends further to the fact that copying here involves the behavior necessary to generate a *sound*, just as the foregoing discussion suggests should be possible, since the son can record its own singing "image," in addition to its parent's, in its outer stream, and, as a playal, will have a record of the neuromuscular activity necessary to yield its version of the song in its inner stream. As explained before, the parent's song will never drive the son's singing directly, but when the two "images" of the song are made to match in the outer stream of the son, subsequent production of the matching behavior is then possible from the record in its own *inner* stream.

COPY-LEARNING AND UNDERSTANDING

We have a *feeling* of relating to components of our environment in a particular way which we refer to as "understanding" them. This raises the question of what might be the basis of this feeling, and what its evolutionary roots might be. It doesn't help with answering the question to say, as dictionaries do, that "understanding" is a manifestation of "knowing," since this just shifts the question of what "understanding" might relate to, to that for "knowing." What we need to do is find some *biological* process with which we can *associate* "understanding" in a convincing way.

As I shall show, *copy-learning* is the biological process we are seeking, but it is worth spending a few paragraphs here to see how we are going to be convinced that some *biological process* can be "associated with"—"corresponds to"—some "feeling." The problem stems from the fact that biological processes proceed at one level, that of the chemistry of molecules, while feelings proceed at a different level: the interaction of different parts of a whole complex brain. We can see that this is so from the fact that whereas we "feel" we understand, we don't "feel" the chemical processes that are at work when individual signals are conveyed from one part of a nerve to another. So we are not going to be able to match *feelings* to *biological processes* with the same degree of demonstrative definiteness with which we could match my right hand with someone else's by simply laying one on the other, and, because right hands are all at the same level, simply *seeing* that they match.

The best we are going to be able to do, in matching *feelings* and *biological processes,* is to rely on what is known as "isomorphism." As it happens, isomorphism is just what the two parts of the word suggest, since they come from the Greek words *isos,* meaning "equal," and *morphe,* meaning "form." So the best we'll be able to do to be convinced that "understanding," the *feeling* proceeding at one level, and copy-learning, the *biological process* proceeding at another level, are the "same," is to show that they are isomorphic. The way we do this is show, first, that "understanding" and "copy-learning" have components or aspects that can be matched one-to-one, even if they do not "look alike," then we show, second, that the aspects or components that match one-to-one behave in the same way. What we aren't going to be able to do, though, is lay the "feeling" of understanding on top of the biological process of copy-learning, and "see" that they match, as with two right hands; for the situation is much more like trying to show that a map which one can hold in one's hands is the map of Canada, for instance.

So it is important to add that isomorphism can, even at its best, only suggest correspondence that isn't already clear without it. In the end, the real confirmation of correspondence comes from what *else* can be explained by the assertion that, when we refer to "understanding," we are referring to the process in a brain by which it captures the behavior of some other entity within its own processes by copy-learning, and so becomes able to reproduce that behavior.

What I shall do now is go directly to demonstrating how all this looks when we try it, in what amounts to a definition of "understanding," as follows:

> "Understanding" is a relation that exists between a creature and particular components of its environment. The relation has two aspects which are usually referred to as "depth" and "breadth" of "understanding." The "depth" of "understanding" is determined by the completeness with which the creature can copy-learn the behavior of a particular component. The breadth of "understanding" of any creature is determined by the number of components in its environment whose behavior it can copy-learn, and this increases as the number increases.
>
> Since copy-learning behavior is a biological process, both the depth and the breadth of understanding, which it is possible for a biologically mature creature to achieve, are determined by its evolutionary status; what a creature can actually achieve is determined by its state of maturity, and increases as its actual state of maturity increases. What it can achieve is also determined by the actual exposure that the creature has had to its environment.

This happens to be a particularly simple isomorphism, unlike some that I shall present later on, but one should not be deceived by its simplicity. Indeed, you will have noticed that there are only two aspects of "understanding," depth and breadth, which can be matched respectively with two aspects of "copy-learning," completeness of copying and number of components that can be copied, and further that the two aspects of "understanding" behave, respectively, like the two aspects of "copying." But it is this matching up between the two sets of aspects of understanding and copying that constitutes the "isomorphism" between them, and that is at the root of the (my) conviction that the "feeling of understanding" is nothing but the "feeling" of the biological process associated with "being able to copy-learn behavior."

Using this view of understanding, what we find in us represents the case in which the breadth of understanding has extended to much of our environment. However, reptiles would not possess understanding since, as I have shown in the previous chapter, there are no playals among them, and hence no ability to copy-learn behavior.

In contrast, playals would begin a succession of evolutionary states of understanding, beginning with shallow understanding in relation to a (female) parent alone. This would be followed by successive evolutionary states in which the playal deepens its understanding of, and broadens it beyond, its parents and siblings. Other kinds of creatures which it encounters, and then finally plants and inanimate entities would subsequently become the subjects of understanding. This final state would correspond to that which we occupy, since we tend to have a feeling of either understanding or being able to understand almost everything around us.

But all this tends to seem rather formal and technical, and perhaps even puzzling, particularly as it relates to copy-learning the behavior of plants and inanimate entities, so it is useful to see, with an example, to what extent such an interpretation of "understanding" corresponds to our everyday experience and use of words. Suppose, then, that I wish to indicate that I "understand" a *piston* (or a polar bear, or a poplar, or a planet). It would be quite natural for me to say: I can show you how a piston *behaves*. What I would then proceed to do is give an account of the *behavior* of a piston as it is stored in my brain, that is, as I have copy-learned it from observing the behavior of a piston. I might proceed to make drawings, or a rough model of a piston, or an elaborate simulation of a piston on a computer, in order to show its behavior, that is, "how it works," but all I would be doing in each of these cases is repeating, to one extent or another, the behavior of the piston as I have cop-learned it into my brain. The only way in which I can demonstrate that I "understand" the piston is to *behave like it*, by using some means or other to enact the behavior that I have *copy-learned*.

Evidently, only a piston can behave fully like a piston, indicating that my (or anyone else's) "depth of understanding" of a piston (or anything else) will always be limited, even though it can be quite considerable. We should also notice that, since the behavior of a piston can have many different aspects, it is possible to have "different understandings," which can differ in both depth and breadth, without any of them being "wrong," in the sense that no aspect of a piston's behavior corresponds to the copied behavior. (If the significance of being able to understand plants and inanimate entities, that is, of being able to "copy-learn their behavior," still seems puzzling, it will seem less so when I discuss this further in Chapter VIII, in the context of "narration.")

This alters somewhat the picture we tend to have of understanding as somehow allowing us to penetrate the understood entity. What the present view is showing is that the understander is more likely to be penetrated than the understood, and that it is the behavior of the *understander* that is affected by the behavior of the understood, at least in the first place, rather than the other way round.

An interesting and quite fundamental question arises as a consequence of asserting, as I have just done, that one can understand what it is "to understand," since, if we have no clear *biological* equivalent of "to understand," there is a strong tendency to "feel" that we might not, or even should not be able to understand what it is "to understand." So the question can be stated as follows: Is there anything about the identification of the biological equivalent of understanding as the copy-learning of behavior in particular that allows one to show that it is possible to understand what it is "to understand"? I believe the answer to this is clearly yes, and the reason is, to put it simply, that one can point to such a thing as "a copy of a copy."

It is important to notice that property of copying, since there seems to be no other behavior like it, in that, while one can point to "a copy of a copy" (and genes made this possible long before the emergence of processes beginning with "X"), this is not possible for "a song of a song," or "a stone of a stone," or, more generally, "a (something) of a ('same' something)." Copying seems to be unique in that it is what one might call "generational," that is, it is the basis of the production of different generations of the "same" entity. So even when we seem to have come on an exception in "a mother of a mother" to whom we can point, we are involved with a manifestation of the "generational" property of copying.

What copying does, then, is allow us to bridge two generations, or, we might say, two "levels" of the "same" entity. So we can begin to see how, if understanding is copy-learning, we can come to be able to bridge two levels of understanding, and point to at least "an understanding of an understanding" which can then be read as "a copy-learning of a copy-learning."

So, returning to the original question, we can see that not only is "understanding" just a manifestation of copy-learning, but only because it is can we come to understand how it is that copy-learning is "understanding." Indeed, it is for this reason alone that we can come to answer the fundamental and completely unavoidable generational question about how one can have an understanding of "understanding."

FEMALE PARENT AS PRIMAL SUBJECT IN COPY-LEARNING

According to the evolutionary process that I am advancing, the parent, and more assuredly, the *female* parent of the playal was the *first entity* to become the subject of copy-learned behavior. As just explained, this can be taken to mean that the female parent of the playal is the first subject that the playal was able to "understand." This was the situation in the context of the emerging evolutionary process as this has extended over thousands of generations, and still is the situation in the process of normal development of the young in later playals, since, under normal conditions, play will always begin with the female parent in mammals, if not in birds.

In view of this, it seems reasonable to surmise that the *biological* development associated with copy-learning might have been such as to reflect a *permanent* need for *starting* with the *female* parent, not just as necessary for the emergence of a new evolutionary sequence, but as *defining* the "normal" course of development of a young playal as this relates to its acquisition of the normal capacity to copy-learn, that is, as a requirement for the normal development of its *capacity to understand.*

If this is the case, then separating a playal from its female parent at birth might produce a disturbance of the *normal* process by which copy-learning and hence *understanding* develops, even if *some* process of copy-learning, due to its hereditary nature, can proceed in the absence of play between female parent and young. But even though *some* copy-learning actually proceeds, one might surmise that the development of a playal which had suffered this kind of early disturbance in the development of the process would continue to reveal *permanent defects in its normal behavior*, reflecting this disturbance in the early development of *its ability to understand.*

There is a great deal of evidence to show that observable behavioral disturbance is associated with such early separation, and, to illustrate this, I shall cite a passage from *Behavioral Neurology* by Pincus and Tucker:

> "...Infant monkeys were taken away from their mothers and 'raised' by surrogate mothers (rag dolls). Although they were healthy and well developed, showing no sign of abnormality in solving problems, these monkeys were unable to establish normal heterosexual relations even after transfer to a colony of normal monkeys. *Partial and total social isolation during the first six months of life also led to severe, irreversible abnormalities in the adult monkey's behavior.* The actual neurophysiological and neurochemical correlates of this state are not known. Isolation does not produce such abnormalities in older monkeys who have been raised by a normal mother."

(My emphasis) (Pincus and Tucker, page viii.)

Infant monkeys are infant playals. So it is interesting that what is, on the surface, such a baffling link, between early separation from a female parent and adult behavior, should be suggested by the model I am advancing in connection with "understanding" in playals.

COPY-LEARNING AND SELFNESS

There can be no doubt that we entertain, as a part of being "normal," a certain feeling of wholeness, combined with a feeling of detachment from the world around us, and of acting from *inside* us, which we sum up in the notion of being, or of having a "self." Furthermore, this "self" is usually endowed with a "personality." Not only do we entertain such feelings, but we tend to imagine that this "selfness" is a singular manifestation of our species, not to be found in other kinds of creatures. Thus, it is of interest to see what part the emergence of copy-learning in the playal might have had in the evolution of this selfness.

A place to begin is with the reminder that the playal was the first (and only) creature to begin to store inner streams *permanently* and to use them in executing copy-learning. Now, in a playal that stores inner streams, the outer streams can be seen as containing two basically different groups of images (where I am still using "image" in the most general sense). One group of outer images depicts behavior that has no relation to the playal's stored inner streams, because there are not even fragmentary images of the playal itself in them. The second group of outer images, depicting the playal itself to some extent, displays behavior that *matches* its stored inner streams in many details.

We have now come on the basis of another isomorphism which, on this occasion, is what is needed to account for the "feeling of selfness," and I shall proceed to demonstrate this in what amounts to a definition of selfness as follows:

> "Selfness" is a relation between a playal and the *totality* of its environment. It has *five* aspects which we might refer to as "*separateness*," "*internality*," "*wholeness*," "*identity*," and "*personality*." "Separateness" is determined by the *dual* existence in the playal of images in outer streams which *match* inner streams, and of images in outer streams which *do not match* inner streams. "Internality" is determined by the fact that the matching which gives rise to separateness is matching between images of the playal in its outer streams and *inner* streams. "Wholeness" is determined by the *completeness* with which the images that match portray the behavior of the playal. "Identity" is determined by the extent to which an outer image that matches and is stored in *one* episode matches that obtained in a *subsequent* episode. "Personality" is determined by the totality of the stored

outer images that match, together with their matching inner images, which form the basis of the copy-learned behavior of the creature.

Since all the processes involved in the isomorphism are biological processes, what is possible for a mature playal to achieve with respect to each of the five aspects of selfness will be determined by its evolutionary status. What any playal can actually achieve will be determined by its actual state of maturity, and the actual exposure that it has had to its environment.

What this suggests is that "felt" selfness emerged first in the playal, at the same time that it began to display copy-learning and understanding. It seems reasonable that "separateness" and "internality" would have been its first manifestations, followed by the addition of "wholeness" and "identity." "Personality" would become more and more complex as the range of copy-learned behavior increases, enlarging the number of inner streams that match outer streams.

It would also seem reasonable that the expansion of copy-learning to a widening range of behavior must have given rise to a further significant development in the nature of the "feeling of selfness" since, as we can see from the definition, selfness must have begun as what we might call "total" or, better, "undifferentiated" selfness, by which I mean that, when selfness began to emerge, and with the exception of the female parent, everything that wasn't the creature itself had the same status of non-self. But, with the gradual expansion of copy-learning, this must have changed, since, as expansion occurred, there were no longer only two groups of images: those in the outer stream that match inner streams, and all the rest, those in the outer stream that don't. With the expansion of copy-learning, there are increasing numbers of images in the outer stream that form a third group in which images almost match inner streams, since this is precisely the effect that copy-learning produces.

Thus, the expansion of coy-learning, that is, of understanding, leads to the emergence of a *differentiated* form of selfness in which the previously undifferentiated selfness increasingly comes to form just a background against which a new kind of "selfness with respect to" particular components in the environment begins to develop, producing, in this way, a new and expanding aspect of the playal "self," built up from the increasing number of components of the environment whose behavior it copy-learns and so comes to understand.

What seems clear, then, is that both "felt" selfness and understanding can be placed in a linked and reasonable evolutionary context in which the emergence of both of them must have preceded that of humans. Indeed, their emergence must have preceded that of humans by the many hundreds of thousands of generations required to go from the first copy-learning playals to ourselves. The emergence of differentiated playal selfness would, evidently, have coincided with the expansion of copy-learning and understanding.

It is interesting to notice that "felt" selfness, which emerges with copy-learning, could not be the first form of selfness to be found in creatures. One might surmise that this should be so since the whole basis of evolutionary theory rests on behavior that is self-serving, and so there must have existed, prior to the emergence of copy-learning, some biological process that also served to define a self, even though this need not have been a "felt" self, that is, a self associated with "feelings."

One such biological system defining selfness is that associated with the "immune" system. This system is the grouping of biological processes that lead to the production of "antibodies" which distinguish between, on the one hand, entities that should form part of an "individual" creature, as defined by its genes (and aspects of its past), and, on the other hand, entities that should not. The entities that should not form part of the "individual" creature, among which some are known as "antigens," make up the entire domain of the non-self. It is such an "immune" system that protects individual creatures, located beyond a certain stage of evolutionary development, against invasion by bacteria and viruses, for example, and which, by the same "logic," rejects grafts of organs, for instance, originating in another "self."

It is worth noting here that selfness associated with both copy-learning and the immune system depends on demonstrations of matching coupled with non-matching, and we can see that, in a very general sense, this will always be the case, since self is the complement of non-self, that is, self plus non-self adds up to "everything," and so the demonstration of selfness by matching will always have to be accompanied by that of non-selfness through non-matching.

We can begin to see from all of this that, starting with the kind of "felt" selfness that I am asserting emerged with behavior based on copy-learning, one can go backward in evolutionary order and find an older kind of selfness in the immune system. One is then driven to wonder whether there might not be a basic evolutionary requirement that, whenever a new way of producing behavior emerges, a new, related form of selfness must emerge with it, since, without such a new form of selfness, a "selfless" form of behavior would emerge, and the creature carrying such "selfless" behavior would soon fall victim to the forces of natural selection, if not to simple suicide.

In such a context, one wonders what the new form of behavior might have been that would have led to the emergence of the immune system. This could have been behavior associated with the emergence of the central nervous system, embodying a brain, since this would clearly have led to an entirely new "pervasive" way of producing behavior, which might be well supported by an immune system having a distinctly "pervasive" quality as well. This would place the timing of the evolutionary emergence of the immune system somewhere between the emergence of the brain and that of vertebrates, which, as it happens, is not unreasonable. Also, it suggests that the central nervous and immune systems should constitute such an evolutionary whole as would account for the increasing evidence of direct interactions between immune functions and activities proceeding in the playal brain.

One can, of course, go back further, to the preceding new way of producing behavior, and one would, presumably, have to identify the single nerve, which provided a new and relatively very rapid way of comparing what was going on outside a group of cells with what genes require to be going on inside, so the group can continue to survive. We can see that, here again, matching is involved in the emergence of what can be seen as the basis of another kind of selfness, relying again on the difference between a stable "inside" and an outside. The new kind of behavior associated with this form of selfness would seem to be withdrawal, in the primitive form of contraction. We can go even further backward in this way, to a single cell, which expresses its selfness by selectively extracting substances from its environment that match other substances which form part of its individual biological constitution, and by rejecting substances that don't match; a process that, evidently, is still at work in us.

In this way we can come to see that selfness is a manifestation of evolutionary persistence, and that all species, to have any possibility of survival, must carry, as part of their biological endowment, all those forms of selfness that are appropriate to the various ways and combinations of ways in which they can produce behavior.

So what we see in late-emerging creatures like ourselves is an overlay of the selfness systems that have evolved. But all of them are not necessarily active in us, since the various "new" ways of producing behavior that we embody might have made the advantages of the simultaneous presence of all the related "selfnesses" not worth the biological cost, and so some might simply have disappeared from the biology of our species, under the pressure of mutations and natural selection.

It would therefore appear that the selfness we "feel" is simply a concomitant of the new way of producing behavior represented by copy-learning. This form of selfness, like all other forms that are active in us, is necessary if our species is to survive, and is simply one member in a long evolutionary series that includes selfness in a number of different forms as a fundamental part of survival itself.

We can now combine this discussion of selfnesses with the preceding discussion of the "female parent as primal subject in copy-learning," and notice that since a newborn playal, and, more clearly evident, a newborn mammal must have shared the selfnesses of its female parent prior to birth, it is not surprising that the first of its copy-learning activities, which determine components of its differentiated "felt" self, should involve this parent, for this is part of the process by which a playal, soon after separation from the biological system of its female parent, must come to establish its own independent array of selfnesses, free of those of its female parent, in order to confirm itself as a separate creature. This helps to explain the disruption in some patterns of behavior, mentioned earlier, which follows separation of primate newborn from female parents, since, according to the suggestion just made, this would defer and maybe ultimately prevent the development in the creature of its own, individual selfness-with-respect-to its female parent, and perhaps with respect to the "femaleness" typical of its species more generally, for instance.

COPY-LEARNING AND PARTITIONING OF THE BRAIN

We need only think a little of the additions that must be required in the brain of a playal, for it to achieve copy-learning, to realize that such additions must have been accompanied by the development of a new segment of the brain in which to lodge them.

This new segment of the brain must be separate enough to allow the new processes to proceed without bringing confusion to the instinctive functioning which preceded them, since, especially at the beginning, instinctive functions must continue to meet the major part of the playal's survival needs. In a later chapter I shall cite some of the very ample experimental evidence in support of the existence of such a separate brain-space.

I shall frequently refer to this new segment of the playal brain as its "learned-routine space," to distinguish it from its more primitive (pre-playal) instinctive space; and the word "learned" is used to remind us that there are activities that take place in this new space which do not have the same kind of relation to the "real world" of the playal as the activities which take place in its previously existing instinctive brain-space.

Evidently, with the emergence of the learned-routine space, the playal begins to have two locations in its brain from which some aspects of its behavior can be determined. One location, the more ancient, is associated with its instinctive routines, while the other, the more recent, is associated with those routines that flow out of the capacity to copy-learn the behavior of components of its environment. This raises a number of important possibilities which I shall mention briefly.

The first of these relates to the possibility, which the playal begins to present, of being able to copy-learn the *instinctive* behavior of creatures of its own species into its learned-routine space and thus eventually circumvent the need for carrying at least some aspects of its behavior as instinctive routines in the form of genetic information, replacing this by a process of copy-learning behavior between one generation and the next, that is, by cultural transmission. Precarious as this might seem, it has the enormous survival advantage of being able to respond to sudden environmental changes which would threaten the survival of creatures shackled to the long period required for reliable behavioral adjustment to occur as genetically controlled change.

This raises the question of how, given two ways of arriving at the same behavior, the playal will avoid states of potentially fatal indecision while it oscillates between the two possibilities. This would seem to be easily resolved in the (evolutionary) short term by the process of natural selection leading to priority of behavior driven from the learned-routine space. At least for some behavioral routines, the question is resolved in the (evolutionary) long term by the gradual loss of the capacity to execute them instinctively due to the disappearance of the biological components which support them as instinctive routines.

The tendency for natural selection to select so definitively for culturally-transmitted rather than instinctive behavior, for some limited number of routines, is simply an expression of the survival advantages of the one kind of behavior compared to the other, in the environmental circumstances of a particular species of playal in which such selection occurs, coupled with the pressure for space in a brain of growing but necessarily limited size.

A related possibility is that, as more and more of the behavior of the playal is driven from copies of behavior, it will come to understand more and more of its own behavior, since, as explained before, this is precisely the nature of understanding. Further to this are the differences between processes associated with instinctive behavior and those associated with behavior driven from copies in the learned-routine space, all of which can be associated with emerging feelings of "freedom" and "will" which correspond to emerging biological features of the creature that are links to the differentiated self discussed earlier.

EPISODIC NATURE OF COPY-LEARNING

It is clear that the whole basis of copy-learning resides in the acquisition of behavioral routines which are derived from encounters that take the form of a short series of linked events which have a story-like character, that is, *episodes* which are short and decisive. This suggests that what I have referred to as the "learned-routine space" in the brain of the playal is stocked with such episodic streams, and that they constitute the elements of "understanding."

If this is so, and I shall give more reasons later on for believing that it is, then we can begin to see, even at this early stage in the development of the playal brain, the origins of the limitations we face, in being forced, as we would seem to be forced, to view the world around us as necessarily made up of episodes, which always have *beginnings* and *endings*, if it is to be "understandable" at all. As is shown in Chapter IX, this has fundamental significance for the emergence of "causes" and "effects" in the functioning of a brain, as well as dominant aspects of the structure of science.

THE LEARNED-ROUTINE SPACE AND CHALLENGE

Evidently, the dominant state of the young playal during play is that of *defense*. That is, the situation in which the playal begins to evolve the development of the stocking procedures for its learned-routine space is one in which it is pervaded by a state of *challenge*. This is clearly a biological state, quite apart from what might be taking place while the playal occupies this state.

It would therefore not be surprising if, because of this, those biological mechanisms in the playal, which allow connections, from both outside and inside it, to its learned-routine space, should become *permanently* linked to the need for a state of challenge. That is, it would not be surprising if any later development, in which image-sequences other than those arising in primal play, but which could also be lodged beneficially for the playal in its learned-routine space, would be accompanied by a requirement that the playal, by whatever means, assume the biological state associated with challenge before the necessary biological paths could be established between the nervous streams from outside and inside it at one end, and its learned-routine space at the other.

I shall, in everything that follows, assume that such a link exists between access to the playal's learned-routine space and the biological conditions associated with challenge, and that, only if the biological conditions associated with challenge are met, can connections to the playal's learned-routine space be established. This assumption, as you will see in later chapters, has very wide-ranging consequences, since it becomes the evolutionary basis for what we refer to as the "unconscious activity" of a brain.

EXTERNAL MANIPULATION AND COPY-LEARNING

By "manipulation" I wish to convey those efforts by one creature (almost always a human) that induce some kind of behavior in another. It is useful, at this point, to sharpen our notions of instinctive and copy-learned behavior by seeing what manipulation can accomplish in the one case and in the other.

In the case of instinctive behavior, an animal can be manipulated so as to induce it to *disclose, from within itself,* various modes of behavior. Thus, the behavior that the animal eventually performs need not and generally will not *resemble* the form that manipulation takes; and what the animal discloses, in the presence of such manipulation, is the range of behavior that its genetic endowment can generate directly as behavior. As the range of manipulation is enlarged, the limits of the range of behavior that can be generated directly by genetic endowment can be mapped out.

The situation is different in the case of copy-learned behavior, for here the behavior that is to be produced *must form part of the manipulation itself,* so that the behavior the animal eventually performs will *resemble* at least a part of the manipulation. Thus, the performance of copy-learned behavior does not disclose what the genetic endowment can generate directly as behavior, but discloses what the genetic endowment can *support* by way of copy-learning itself.

So, in summary, manipulation leading to instinctive behavior induces the disclosure of what the animal has stored within it that allows it to execute, as external performance, behavior that originated *inside* it; while manipulation leading to copy-learned behavior induces the disclosure of what the animal has stored within it that allows it to *copy* and execute, as external performance, behavior that originated *outside it.*

Now, given the role that I am asserting proprioception must be playing in copy-learning (and which much of the rest of the book will confirm), we come on the interesting (and contentious) question as to whether external manipulation of limbs by a human therapist (which would, evidently, produce an inner stream by stimulating proprioceptors), could assist with the process of copy-learning the "behavior" present in the manipulation. The foregoing discussion certainly suggests that it could, and even suggests the minimum conditions, which would be that the intended behavior must be part of both the inner and outer streams, that is, that *some image of the manipulation must accompany the manipulation itself.* This should not be the basis for a wave of quackery, being only surmise. But one does not easily discount the many generations of teachers who have held and guided the hands of children learning to write, while young brains collected images of the magic flowing off the ends of pencils. Quackery?

Section C: Uniqueness of Playal's Situation

UNIQUENESS OF PLAYAL AS HOST FOR COPY-LEARNING

The main theme of this book is set by my attempt to show that play was an evolutionary phenomenon which, because of its own particular features, provided a unique situation in which the playal became the host for a series of evolutionary developments which led to what we have become. Some of this has been shown in the discussions in the preceding chapter and, even more particularly, in the present one. But since showing this uniqueness is the dominant theme of the book, it would seem useful to summarize the salient points already mentioned, and add any other confirming observations that can be made at this time.

I begin by recalling the requirement to thoroughly isolate the image of the first actor, in the outer stream of the playal, whose behavior is to become the subject for copy-learning. The female parent in play becomes the ideal subject, since, while the young playal is under attack in play, its instinctive program focuses the whole of its attention on the field occupied by its single attacker alone. As a result, the situation of play provides, in a powerful and automatic way, precisely the conditions required for satisfying this requirement.

Furthermore, it is only when the attacker is the parent of (the same species as) the young playal that what it might begin by copying is sufficiently close to possibilities in the playal itself to provide the basis for the very first instance of copy-learning. Coupled with this is the freedom of the playal to carry out harmless repetition in which the subject whose behavior is to be copied remains the same from day to day, thus making the situation of play entirely different from anything that could prevail in the context of random attacks by creatures other than the parent of the playal.

Finally, there is the continuing convergence of the process of copying, provided directly by play, and which results from the fact that, the closer the behavior of the playal gets to that of the parent, the closer the copying process comes to being realized, and, in turn, the closer the process comes to being realized, the closer the behavior of the playal gets to that of the parent, and so on, until a copy is realized.

What all of this suggests is that, even if no single one of these aspects of play had been unique, the coming together of all of them must certainly have constituted a unique and remarkable evolutionary conjuncture.

The particular aspects of uniqueness just discussed derive from a generally unique relationship between the young playal and the restricted environment in which it must begin to acquire its procedures for realizing copy-learning. But it is also important to see to what extent uniqueness might exist in the relation between the young playal and the wider environment in which it must beneficially apply copy-learning.

UNIQUENESS OF PLAYAL FOR APPLYING COPY-LEARNING

We need to look at the relationship between the playal and the environment in which it must apply coy-learning because, even if the particular conditions found in play are unique and ideal for initiating the emergence of copy-learning, the modifications required in the brain of the playal are so extensive that they would probably require hundreds of thousands of generations to evolve.

The pressures of natural selection would require that, during all that time, emerging copy-learning bring consistent benefit to the playal, in spite of all the changes in the environment which could be expected to intervene over such a time. Put differently, the biological changes that are slowly taking place must be such as to increase the capacity of the playal to survive major environmental changes, or they will almost certainly lead to extinction at some point in this long chain. Only improvements in the capacity of the playal to hunt and to defend itself could meet such a demand on its biology, and so it is useful to look at the extent to which the acquisition of copy-learning affects these aspects of its behavior.

It is not difficult to see that the ability of the playal to copy the offensive behavior of its parent leads directly and consistently to an improved capacity for it to hunt and kill its prey and would lead to an improved capacity for it to defend itself as well. What is worth emphasizing here is the fact that the improvements will tend to be transmitted directly from parent to young via play, rather than by the long and not so reliable process of hereditary transmission. Thus, when a young playal begins killing on its own behalf, its behavior will be close to that of a mature parent, complete with all the latest successful variations, which would simply form a natural part of the parent's attacks in play, and be copied by the young. This must clearly increase the capacity of this new emerging type of playal to withstand even severe changes in its environment, and improve its capacity to survive the ever-shifting pressures of natural selection.

Furthermore, not only does the playal gain directly by copying the behavior of its parent, but it also gains indirectly because copying the behavior of the parent, in particular, makes it possible for the behavior of the young playal to approach that of its parent earlier in life, with the result that mutations in the young, which lead to differences in behavior between parent and young, show up more clearly and sooner, and so the beneficial ones have a longer period in which to act and confer their benefits.

So far, I have been discussing the uniqueness of playals as a locus for the emergence of copy-learning by comparing them tacitly with all other animal categories. It is useful, as the last part of this section on the uniqueness of the play environment, to compare playals directly with plant-eaters.

UNIQUENESS OF PLAYAL COMPARED TO PLANT-EATER

Even though it seems that plant-eating animals can be viewed as tending to be outside the stream of play, it is useful, for reasons of contrast, to consider them here and to see how they would relate to what has just been said, if they did play in the sense of wrestling and biting in a muted, artificial form of killing as in the case of "true" playals.

The point I wish to make is that such play in plant-eating animals would not lead to learning anything that relates to their survival in the real world of their food-gathering needs. Thus, animal-eating play in plant-eaters, even if it did occur, could not lead to any major, sustained evolutionary modification in them, of the kind I associate with playals, simply because there would not be sufficient real-world survival benefits attached to what they could learn in such play to sustain a long evolutionary process which progresses only at a high biological cost.

But if, rather than looking at what hasn't evolved, we look at what has in plant-eaters, what we find is mainly solo and group exercise activity, including chasing and butting, as with the cattle of the Camargue referred to previously. This could be interpreted as having significance for general physical conditioning, as well, perhaps, as some preparation of the young in defense based on flight. But what this play-like activity lacks, and this is crucial when compared with that in the "true" playal, is a direct link to real-life food-gathering.

It is therefore not difficult to see the substantial relative advantage that the "true" playal possesses as a site in which copy-learning would emerge and continue to develop. The evident power of such creatures which, to a greater or lesser extent, embody this world of the artificial in a separate part of their brains, is achieved only by the fortuitous coming together in them of the unique array of features that, in primal play, reveal and resolve the paradox of their existence as animal-eating animals which rear their young in close proximity to themselves.

CHAPTER IV
S L E E P

Why Discuss Sleep?

As you will have noticed, much of the preceding discussion is concerned, in one way or another, with human behavior and its evolution. And it was not so very long ago that such a discussion could have continued without even a mention of sleep. But the situation has changed markedly since the early 1950's, when researchers like Eugene Aserinsky, William Dement and Michel Jouvet began to call attention to the behavioral structure of sleep, and the way in which dreams form part of it. In the intervening fifty or so years, an enormous amount of work has been done, by them and hundreds of others, which shows quite clearly that the third of our lives occupied by sleep is not spent in biologically irrelevant inaction.

However, there still is no theory that can account satisfactorily for, and bring together, the main features of what is now known of the behavior that accompanies sleep. So it is interesting to see if what is known of sleep and dreaming can be positioned within the play-derived behavioral developments I have been discussing, and if this positioning can help to explain what is observed. In order to do this, I shall, in effect, be trying to develop an evolutionary explanation of the emergence of sleep.

(This is a long chapter, which touches a substantial number of the aspects of sleep. So I urge you to make the considerable effort needed to follow the discussion, and to share my view of sleep as a remarkable crossroads in the life of a brain, which, when we stumble on ways of reading its signposts, can lead us along many interesting paths.)

The Emergence of Sleep

To see how sleep might have emerged, let us start with a creature that precedes reptiles and playals, but that has a nervous system with a brain which exercises central control of some significant portion of its behavior. All vertebrates have such a nervous system, but even some of the more ancient creatures (arthropods), such as insects, crabs and other crustaceans, as well as spiders and scorpions, do also.

As indicated before, such creatures must have three basic kinds of nervous sub-system. The first, its outer stream, must be able to tell enough about the creature's environment to allow it to gather food and defend itself. The second must allow it to control its movements by exciting muscles that are remote from the brain, and, in a creature at the evolutionary stage being considered, will be an instinctive (genetically controlled) sub-system that drives instinctive behavior. The third sub-system, relying on proprioceptors, must provide the creature's brain with a sufficiently accurate and continuous indication of its configuration to allow the movements being demanded by the instinctive sub-system to relate to the real possibilities of the creature at any time. In summary, then, there are *three* nervous sub-systems: a first one, generating an *outer stream*, which reports on the environment; a second one, generating *instinctive behavior*; a third one, generating an *inner stream*, which reports the state of all the parts the central brain controls (and which, much later in playals, becomes the basis of copy-learning, as explained in the previous chapter). Let us now focus on the portion of the brain that is fed by this inner stream, since it is here, as I intend to show, that the origin of sleep is to be found.

It is not difficult to imagine that the inner stream will flow into some store in a portion of the brain in which the creature's configuration is represented. Now, sooner or later, *errors* are sure to accumulate in this store, and, regardless of the details of the way in which the configuration represented in it comes to depart significantly from the creature's actual configuration (and I shall return to some of these details later in this chapter), some way must be available for allowing the store to free itself of the errors, and then "catch up" with the actual configuration. This is so because, if an evolutionary excursion should lead to a creature without a way of erasing such discrepancies between its actual configuration and the representation of it in its brain, the creature could not survive for long, since it will continue to make demands on muscles that will respond increasingly inaccurately due to the differences between their actual states and the representations of these states.

Evidently, as the errors accumulate, they entail a growing degradation of the behavioral routines available to the creature, and place it in a condition of continually growing "stress," as this condition was defined in Chapter I. So there must be a way for the creature to clear errors from its inner-stream brain-store and start a new wave of storage in which the configuration data is more current. This can happen if the growing stress induces a state of "quiescence" in the creature.

Now no reasonable evolutionary argument would present the creature as "setting out to resolve its difficulty" in this way, but if, during such a quiescence, which is brought on simply by the mounting stress, the nervous sub-system (inner steam) involved in actually driving the configuration store, as well as that involved in directing muscles (instinctive behavior), *ceased operation temporarily* simply as genetically originated accidents associated with the onset of stress, the inner-stream store could discharge and prepare to receive a new wave of current configuration data. Note that *both* sub-systems need to cease operation, since the inner-stream store must cease receiving data so it can recover, and the sub-system driving instinctive behavior needs a current version of the configuration data so as not to create still more stress, and possibly even destroy its host.

Although this genetic accident would solve the problem of configuration updating of a central brain, and could be operated on by natural selection, it is important to notice that it does so at the expense of an *endless series of periods of quiescence* which interrupt the creature's full functioning. These inescapable periods of quiescence that characterize the functioning of a creature with a brain having central control are, I believe, the earliest evolutionary manifestations of what we come to know as "sleep." When the creature is not in one of these periods, we say that it is "awake," by which we evidently mean: functioning with all of its three nervous sub-systems and its configuration brain-store operating.

In all this, we can see a quite reasonable evolutionary context in which to place the emergence of sleep. What becomes clear from this way of imagining its emergence is the extent to which, far from being a more or less useless activity, sleep is a biological process without which the central brain could not have emerged in even its simplest forms. Another way of viewing this, which is crucial for the evolutionary case, comes from noticing that a consequence of the process just outlined is that sleep had no *purpose* as such; it was simply a biological necessity, satisfied by a biological accident that was processed by natural selection, and which accompanied a central brain as an inevitable part of its evolutionary emergence

Further, since it becomes clear that the initiation of sleep must have formed part of the activity of the most primitive form of brain, then, even in a brain as highly evolved as ours, the initiation of sleep, and at least some of its ongoing mediation must be associated with its most primitive part (brainstem). Although this conclusion comes quite readily out of the kind of evolutionary origin of sleep being advanced here, it has taken a great deal of experimentation with actual brains to demonstrate this connection between sleep and the most primitive parts of a brain.

Evidently, we can now turn the argument around, and say that, since it is now known from experiments that the mediation of sleep is associated with the primitive part of a brain, only a theory of sleep that has such a connection built permanently into it could survive the endless falsifibiality testing that any good theory must. This reduces quite drastically the number of theories of sleep that one can usefully entertain. In particular, it eliminates all those theories that are based on biological processes that appear, only for the first time, as late as in reptiles, or, even more so, in mammals, since such theories would not have the essential evolutionary link to the primitive part of a brain that experimental evidence now requires. So it is far from trivial that the theory being advanced here meets such an evolutionary constraint, and this increases confidence in the assumptions that underpin it.

We can summarize these underpinnings by saying that the need for sleep is prompted by the stress flowing from a lack of current *information*, not a lack of "energy." So a single sleep event should suffice to return the creature to its normal waking state, since this should suffice to allow clearing the errors from the configuration store, which is all that is necessary to overcome the need for sleep.

The Emergence of REM Sleep

As anyone familiar with the literature on sleep will know, what I have been addressing is scarcely half the story, since one of the great contributions of the careful modern observations on sleep has been to show that, in groups of animals beyond reptiles, sleep is not the uniform state of quiescence that the foregoing discussion would suggest. What this work shows is that, at this stage, sleep (which I take to be) of the kind that I have been describing is interrupted intermittently by passages of paralysis and dreaming under conditions in the brain that resemble the *waking* state.

This kind of activity during sleep was sufficiently striking and surprising when first observed that it was quickly named "paradoxical" sleep. It was also named "REM" sleep by William Dement, because it is accompanied by Rapid Eye Movements. It has also become common to represent the Non-REM parts of such sleep as "NREM" sleep, and I shall use this for the kind of primal sleep that I have been describing, since, as will be shown shortly, this is almost certainly what it is.

Thus, using the standard terminology, I consider that a basis for the emergence of NREM sleep has been developed. What I shall do now is complete the outline of the theory so as to include REM sleep in creatures that will be assumed, for now, to be playals.

It is convenient to begin by constructing a diagram of the process for NREM sleep as just described, and then to follow this with two other diagrams which lead to the process by which REM sleep emerges out of that earlier form of sleep (and complicates it). The diagram for NREM sleep is shown in Figure IV.1, on page 60. To orient yourself, begin with the axis OX, which represents the passage of time. The time from TM to TM' is the period of a full cycle of waking and sleeping.

So imagine that TM is the time at which the creature has just awakened from a previous passage of sleep. The period TM-TS, which is the time during which the creature is awake, is also the time taken for the data in its configuration store to lag so far behind its actual configuration that the associated stress induces the onset of sleep. This increasing lack of correspondence is shown by the line M-S, where S is the point at which the critical level of stress has been reached; so the vertical axis of the diagram in the section TM-TS represents stress to some scale, the actual value of which is of no importance for the present discussion. When the critical level of stress S is reached, the process leading to sleep is set in motion.

This process is basically one in which the intensity of the critical stress leads (I assume) to the production of what must be a relatively complicated chemical substance in the brain, which has the effect of suppressing external behavior by attenuating the flow of signals which trigger instinctive routines, and by blocking the inflow of inner-stream data into the brain via the portion of the (afferent) subsystem that transmits configuration data arising in proprioceptors. Both these blockages can be realized inside the brain itself.

As more of the blocking-substance is produced, and its effect becomes more pronounced, the creature progresses further and further into the depths of sleep. This is shown in the diagram as the curve S'-N, and so the vertical axis in the period TS-TM' should be taken to represent, to some scale, the effectiveness of the blocking substance in both immobilizing the creature and arresting the flow of inner-stream nervous signals. Production of the blocking substance can be taken to cease before T12, prior to its effectiveness attaining the maximum level N, at time TN. This corresponds to the deepest level of sleep that the creature will attain, and, as the diagram indicates, sleep remains for a time at this deepest level before beginning to wane, as the effectiveness of the blocking substance begins to decrease.

In the creatures to which I am here referring, that is, those in which a central nervous system is just beginning to emerge, and hence in which sleep is just beginning to appear, it terminates simply with a continued decrease of the effect of the blocking substance, as shown by the segment of curve N-M'.

The inner-stream store has had time to clear itself; the flow of inner-stream signals is resumed; the creature's instinctive routines begin to respond to its environment again, and it returns to the waking state at about TM' , as sleep slips gradually away.

Evidently, changes must be occurring in the creature during sleep, and it is essential to see, in some detail, what these might be. Thus, as shown in Figure IV.2, on page 62, I have divided up the vertical axis between S' and N into four "stages," labeled W, 1, 2, 3. Below the diagram a table has been added which summarizes what is happening during these stages, as the creature moves into and out of sleep. The table shows the states of the three

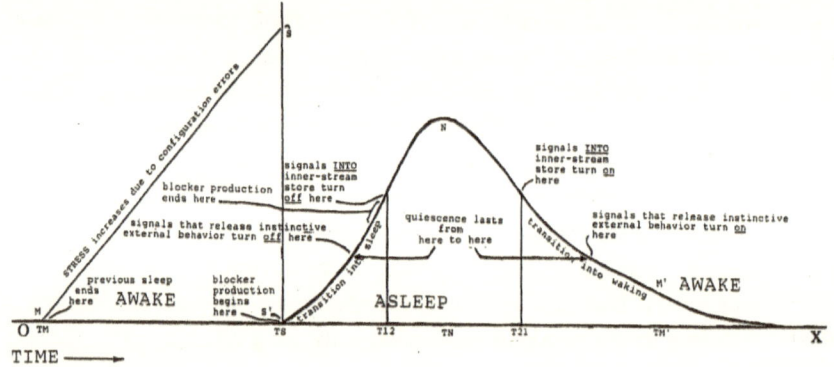

FIGURE IV.1

PRE-PLAYAL SLEEP

signals that are involved as either ON or OFF, although the actual changes of state are doubtless more gradual than this. The outer stream is always ON, that is, sensory information from outside the creature can still flow into its brain. The stages are reasonable in themselves, but I have also chosen them to begin to approximate the generally accepted gradation of stages surrounding REM sleep, to which I shall now turn.

(I take this opportunity to say that, although the following explanation of REM sleep demands considerable attention and care in reading, it will provide the key to understanding much of what comes after it. If necessary, re-read some of the more demanding parts.)

It is well established that REM sleep is found only in creatures that evolved later than reptiles, and so I shall begin to look at it, for reasons that will become clearer as we proceed, in the context of playals, which form a wide group of post-reptiles. To grasp where the difference originates between the previous kind of sleep and REM sleep, it is essential to recall that the primary difference, between the brain of the kind of earlier reptile and pre-reptile creature that I have just been discussing and that of the playal, is the presence in the playal brain of an *inner-stream store*, that I have called the "learned-routine space." This space, which is new in the playal, has an output side that delivers nervous signals which can drive muscles and produce *copy-learned* behavior. So, in addition to the *three* brain sub-systems present in the earlier pre-playal case (a first one, generating an *outer stream*, which reports on the environment; a second one, generating *instinctive behavior*; a third one, generating an *inner stream*, which reports the state of all the parts controlled by the central brain), in the case of the playal there is a *fourth* brain sub-system, generating nervous signals that can drive muscles and produce *copy-learned* behavior, and which I have called the "learned-routine space."

This (and here I repeat, because of its importance) is the space in which, as explained in the previous chapter, all the behavioral routines learned by copy-learning are stored, and these, as you might recall, are built up from permanently stored inner streams of episodes which allow the playal to copy and reproduce the behavior of first its female parent, and subsequently whatever else it comes to copy, that is, whatever else it comes to understand. Thus, in addition to instinctive behavioral routines, as possessed by the creatures that preceded it, the playal has its copy-learned, culturally-transmitted routines as well, which are stored in its learned-routine space.

If you are unfamiliar with the literature on sleep, you might consider the degree of anxiety and stress in the model to be presented a sure indication of its weakness. So I should say, before going further, that measurements made on rates of breathing and heart-beat, as well as on blood pressure and electrical activity in the brain and elsewhere, indicate that the periods of REM sleep are always times of great nervous storms, accompanied by deep paralysis of the muscles that drive external behavior. Thus, rather than being weak, only a model that leads one to expect such manifestations of extreme stress can hope to form the basis of an acceptable theory of REM sleep which, evidently, must not only include these manifestations, but also explain how they come to be.

So let us go now to Figures IV.3a and b, on pages 96 and 97, and see what we can make of REM sleep. Notice that the left part of Figure IV.3a resembles that of Figure IV.2, but that a heavy-black line ("path of playal sleep") has been added. Further, the "SIGNALS" box at the left of the table at the bottom has *four* items in it (not three, as in Figure IV.2), since it must now include the additional signals from the *learned-routine space*. Then, below the SIGNALS part of the table, there is a row entitled PARALYSIS, and you will see in a moment the extremely important role this plays in the explanation of

REM sleep. In addition, I have increased the number of stages on the vertical axis to five: W, 1, 2, 3, 4.

To get started, imagine that the playal moves toward sleep at time TS, for the same reason as did the earlier creature, that is, the need to allow its inner-stream configuration store to clear and restart without errors.

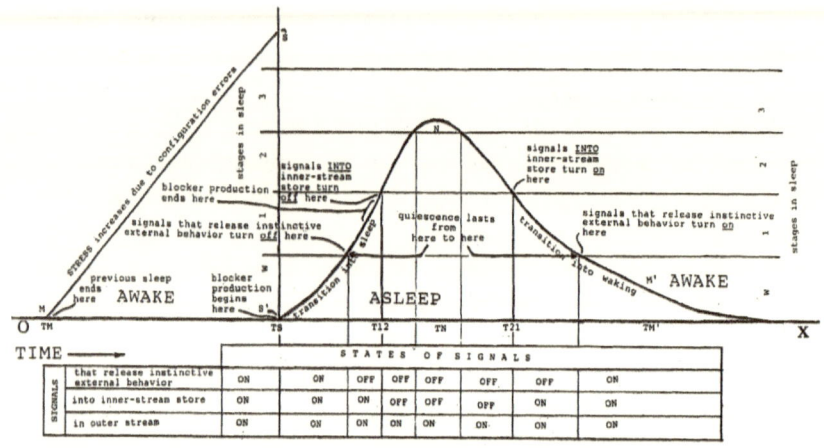

FIGURE IV.2

STAGES IN PRE-PLAYAL SLEEP.

And here we come on the first real challenge in constructing a model of REM sleep, for, if the inner-stream in the playal, which carries configuration information, is stored permanently in its learned-routine space, so it can reproduce copy-learned behavior, as I have been asserting earlier, how can it be assumed here that there is what must be a volatile inner-stream store, which carries configuration representation, as in the reptile, and which is about to clear itself, by initiating sleep? In effect, it appears that I am assuming the inner-stream store to be permanent and volatile all at once! However, as a discussion in a later part of this chapter will show, not only can this apparent inconsistency be dealt with satisfactorily, but, in doing so, it is possible to explain some fundamental features of the playal brain, involving memory. It happens though, that this discussion is quite lengthy, and seems out of place right here. So what I shall do is give a brief summary of enough of it to allow the discussion that continues after the summary to be freed of the apparent inconsistency, even though this has the appearance of letting my story get ahead of itself.

In summary, then, what the later discussion will show is that it is extremely likely that the playal has a permanent inner-stream store that consists of a number of layers of memory. Each layer is created as a new layer during sleep that includes REM activity, and such a new layer can store inner-stream signals until the next REM sleep, at which time the new layer also becomes a part of the permanent memory, while another new layer that can store inner-stream signals is added on to the memory, and so on. Thus, at any time while a playal is awake, its complete inner-stream store consists of two parts, a permanent part which consists of all the old layers, and a new part consisting of a single new layer in which inner-stream signals can be stored until the next period of sleep occurs, at which time this layer also becomes a part of the permanent store, to be replaced by another new layer that develops during sleep.

It is each single new added layer, which stores inner-stream configuration signals that, in the playal, behaves like the single, volatile inner-stream configuration store which is the only store in reptiles, and in which the beginning accumulation of configuration errors leads to blocker production and the inducing of sleep. It is also the inner-stream flow into this new layer of memory that the blocker arrests as a part of the induction of sleep in the playal. Evidently, it will be necessary, as is done in the fuller discussion later on, to show how the single volatile inner-stream configuration memory of the reptile gradually evolved into recurring layers of new memory in the playal. But, for now, without any inconsistency, we can imagine the playal moving toward sleep for the same reason as did the reptile, that is, because of the stress and the release of blocker resulting from the accumulation of configuration errors in a single new layer of its inner-stream memory.

Returning then, to Figures IV.3a and b; as the effects of the blocker intensify, and the playal goes deeper into slumber just after time T12, sleep makes its first departure from that of reptiles, since (I assume that), just after T12, the signals that normally flow *from* the playal's learned-routine space, to muscles, during the waking state, *are not yet turned OFF(are ON)*, *unlike* the signal into the(new layer of) inner-stream store which was turned OFF at T12, and the signal that releases behavior from the instinctive program store, which was turned OFF just before T12. (It is important to follow the sequence being described in the table at the bottom of Figure IV.3, that shows which SIGNAL is ON, which is OFF, and when.)

It is reasonable that the signals from the learned-routine space should be turned OFF *later* than the other two, because the output side of the permanent, learned-routine space, being a *new* location in a playal brain, would not be as ready a target for the blocker as the *older* locations that it had stilled in hundreds of thousands of generations of preceding, simpler brains. Indeed, one can imagine that the signals out of the learned-routine space wouldn't turn OFF at all, unless some new development would emerge from what is, for the sleeping playal, a new situation. So I shall assume that the signals from the learned-routine space are *relatively late* in being turned

OFF, but not simply because of the immediate reasonableness, but because this then allows the entire structure of REM sleep to be explained, as we shall see.

The assumption, then, is that the playal is in a new situation, in which the flow of signals from the learned-routine space to muscles can continue, while the inner-stream flow to the new layer of memory is blocked, and so the signals that continue to flow from the learned-routine space, which are directed at muscles, do not seem to have any "answering" configuration signals back from their proprioceptors, as is normal and essential in the waking state. So this would be a time of great danger for the playal, a new kind of danger, since signals from its learned-routine space will, in the absence of some new development, drive its limbs, for instance, in totally unconstrained ways, for there are no inner-stream signals to guide and limit motion demanded by a learned-routine space running without any indication of where limbs are, and how fast they are moving.

Thus, for the playal to have survived these new dangers of sleep, it must have become the focus of a new evolutionary development as a part of its sleep, and the form that this took was the disabling of the muscles of its motor system which the nervous stream from the learned-routine space would normally drive. How this comes about, we shall see in a moment. This new evolutionary development can therefore be viewed simply as one that allowed the playal to survive the emerging complications of its (new) kind of sleep, by going into a state of paralysis of the muscles that drive its external behavior.

But, of course, evolution never sets out to save any line of creatures, not even playals, and so the question arises as to how, in the uncaring manner of evolution, the state of paralysis comes about, and how it comes to disappear. What would prompt its occurrence in the first place, and, since the playal doesn't remain paralysed forever, its subsequent disappearance?

To answer these questions, we need to begin by noticing that, in a brain equipped with a learned-routine space and the associated attributes of copy-learning and felt selfness, the blocking of proprioceptor (inner-stream) signals to its brain will have as its correlate something like: "my body has disappeared," and such a brain will go into a stressful panic, amid "feelings" such as "floating without a body."

This leads to more frantic bursts of nervous signals from the learned-routine space directed at muscles, and these signals *also seem* to go "unanswered" by proprioceptors, and so these rapidly intensifying signals soon overwhelm the ability of the muscles to accept signals from anywhere in the entire efferent nervous system which would normally cause them to retain their tone and to contract. (Here, I am assuming the onset of a form of what is known technically as a "Wedenski inhibition," but its details need not concern us here.) Such neuromuscular overwhelming, if allowed to disappear by arresting the excessive flow of signals, could last only a few seconds at most, and this would represent the maximum duration of ensuing paralysis.

But the fact that the signals from the learned-routine are now effectively blocked at their interface with muscles, by the overwhelming flow of signals themselves, means that no muscles move, and no inner-stream signals can now even be generated by proprioceptors in answer to what are now the saturating signals still flowing from the learned-routine space. In this way, the panic and the bombardment by signals from the learned-routine space are actually locked in by the resulting saturation of the neuromuscular system itself. So the inner-stream flow is now doubly blocked: first, by the blocker that is inducing sleep, in a process confined to the brain, as it had been in reptiles, and, second, by the saturation of the neuromuscular system due to uncontrolled bombardment of muscles by signals from the learned-routine space, which causes paralysis, in a process that extends outside the brain to the junction between nerves and muscles.

This could continue indefinitely, holding the playal in a state of near permanent paralysis and panic, but the situation lasts only a short time, since (I assume that) such intense stress induces *greater production and spread of the blocker* which is already inducing sleep, and this increase finally *does* succeed in turning off the flood of signals from the learned-routine space, at time Tz. This arrests the flow of the unanswered signals that were originating there; the neuromuscular system recovers in a second or so after the frantic bombardment by unanswered signals ceases, and hence the paralysis disappears; the production of sleep-inducing blocker terminates because the panic and stress disappear; and the playal returns to a course of quiet sleep which resembles that of reptiles again, but with the difference that the increased production of blocker has driven it into a deeper state of sleep at NP, in stage 4, than any reptile had ever experienced.

Evidently, the duration of this entire episode, from the time when it starts, just after T12, to the time of return to quiet sleep at Tz, is completely dependent on the difference between the level, and hence the time, at which the blocker turns off the inner stream at T12, and that at which it finally turns off the flow from the learned-routine store at time Tz, a little later. Indeed, if the two levels, and hence times, are close enough, the episode would go almost unnoticed.

I shall refer to this potentially very short sequence of panic, paralysis and blocker production as "Event-0(zero)." It is not difficult to see that, for a short period at its start, and before paralysis sets in, there might be short bursts of uncontrolled movement, and these are, in my view, what are known as "myoclonias."

There is another and very important aspect of what has just been described that needs close attention, because I have assumed, for what seem acceptable evolutionary reasons in themselves, but which will seem even better reasons shortly, that the intensity of blocker needed to *turn off* the flow from the learned-routine space is *greater* than that required to turn off the flow incident on (the new layer of) its inner-stream store. This implies that, as the effects of the blocker wane, and the playal begins to drift back toward waking, from NP, the signal-flow from the learned-routine space will come back ON again *before* the flow from proprioceptors into the inner-stream store come back ON—more blocker is needed to turn the learned-routine-space output OFF on the way *up*, so the learned-routine-space output turns OFF *later*; whereas, as the blocker wanes, on the way *down*, the learned-routine-space output will come back ON *sooner*. Let us continue, now, with Figures IV.3a and b.

Imagine the sleep of the playal drifting back toward waking from NP, in Stage 4, then the flow of signals incident on the inner-stream store, which went OFF at time T12 when going into sleep, should come back ON at time T21, this being the time at which the effectiveness of the blocker has drifted back to the value it had at T12. However, as just explained, the flow from the learned-routine space will (have to) go back *ON* at point "a," some time *before* T21, which I have identified as Ta. The result of this is that, during a period that runs from Ta to T21, beginning at point "a," we have *another* case of signal flow from the learned-routine space without there *seeming* to be any answering proprioceptor flow, and, as described for Event-0, the playal brain goes into a new state of stressful panic.

As in the situation at Event-0, the "unanswered," uncontrolled flow of nerve-signals from the learned-routine space, which is incident on muscles, quickly leads to their paralysis. The same situation develops, in which the start of the paralysis serves to perpetuate the paralysis itself, since it serves to prevent even the generation of inner-stream signals, because the muscles are unable to respond to the signals incident on them. This keeps the playal in a state of panic, and so on, in a circle that helps to render the inner stream doubly locked OFF, first by the paralysis of muscles, and second by the remaining effect of the sleep-inducing blocker, which is now drifting out of Stage 2, into Stage 1.

Of course, when sleep drifts into Stage 1, at T21, the path for the inner stream does come back ON, but even though the blocker releases the inner-stream path inside the brain of the playal, the continuing paralysis of muscles prevents the generation of inner stream signals by proprioceptors that would "answer" the signals from the learned-routine space, and so, effectively, the flow in the inner stream does *not* come back ON, and the situation continues unchanged, as sleep drifts into Stage 1, closer and closer to waking.

But the process begins to reverse at point r1, because (I assume that) the renewed panic and stress *induce the re-start of blocker production,* and this begins to drive sleep out of Stage 1, back into Stage 2. Finally, the increasing effect of the blocker turns the flow from the learned-routine space OFF at "a'." When this happens, the paralysis ceases quickly; the playal returns to the quiet sleep of reptiles; blocker production ceases in the absence of stress, and the continuing effect of the terminated burst of blocker drives the playal into deep sleep in Stage 4, at point "ab." The process involving panic and paralysis *turns itself off* by inducing the production of a burst of blocker! I shall refer to the period just described which stretches from "a" to "a'" as Event-1.

Having been driven into deep sleep, after Event-1, at point "ab," the playal begins to drift back toward waking, as the effect of the blocker wanes. At point "b" the signals from the learned-routine space come ON again and, since the inner stream is still OFF, the same sequence of panic and paralysis ensues, with the difference that (I assume that) the brain is not able to sustain full production of bursts of blocker without resting for much longer than the time available between these recurring events of panic and stress. Thus, as shown in the figure, because of this growing reduction in capacity to produce blocker, the period of paralysis and dreaming that starts at "b" lasts *longer* than the previous one. Also, when it ends at "b'," sleep is not driven into as deep a stage as before, going only as far as point "bc," in Stage 3. This produces Event-2.

Sleep then begins to drift back from point "bc" toward waking, and, at point "c," the sequence of paralysis and blocker production begins again, and the continuing reduction in the capacity to produce blocker makes this passage *still longer* than the one that preceded it. Enough blocker is finally produced to terminate this passage at "c'," and yield Event-3, but this time the blocker barely succeeds in turning off the learned-routine stream, and drives sleep only as far as point "cd."

From "cd," sleep drifts quickly back to point "d." Here the final passage of panic and paralysis begins, for now the production of blocker is insufficient to turn off the flow from the learned-routine space. And so, after a brief deepening toward the point "de," sleep drifts toward waking at P, as the production and effect of the blocker are exhausted. Just before P, all the functions of the playal resume their normal waking states, except for a remaining paralytic condition which is situated in what has now become its waking context. But it recovers quickly from this, as its brain, now exposed to all its normal inputs, becomes able to distinguish between a missing inner stream and a "missing body." It arrests the output from the learned-routine space just momentarily, which allows the paralysis to cease, and an inner stream to flow from proprioceptors into a new layer of inner-stream store that has developed during sleep. I shall refer to the period of paralysis and waking that stretches from "d" to P as Event-W.

It is interesting to notice that while Event-0 is just a short burst, Events 1 to W last for extended periods of time. This striking and important difference is due entirely to the fact that Event-0 begins at a point in sleep at which the effects of the blocker are *increasing*, and the playal is moving deeper *into* sleep, whereas Events 1 to W begin at points at which the effects of the blocker are *decreasing*, and the playal is moving further *out of* sleep. However, the structure of Event-0 is identical to that of the others, and the difference is that it is quickly snuffed out by the powerful increasing effect of the blocker from a rested source, whereas the later Events, due to the decreasing effect of the blocker and an exhausting source, are allowed to build up and display their full structure, *before* the production of the blocker resumes, and its effect spreads and begins to increase again.

We could reasonably expect that, at the beginning of each of the Events 1 to W, as with Event-0, there would be some movement of the playal before paralysis sets in. It would also be reasonable to expect some "twitching," even during the periods of paralysis, since the blocking due to the saturation of the interfaces between nerves and muscles will not generally be so complete as to prevent all traces of the effects of the bombardment by the signals from the learned-routine stream.

As we have seen, even playal sleep should have time for the previous quiet of the sleep of reptiles, since the state of the brain of the playal, during all the time spent ascending into deeper sleep, after the learned-routine space has been turned off at the end of every Event, and all the time spent drifting back till it turns on again, is almost identical to the state of the brain of the reptile during *its* sleep—neither brain then has an active source of signals suitable for driving muscles. Thus, during these relatively quiet periods, eyes might drift about, but will not execute any sort of controlled, "driven" movement in either creature; the periods *between* the Events in playal sleep will be largely free of eye movement, and certainly free of rapid eye movement. They should, indeed, be NREM periods.

During each of the Events 0 to W, a playal continues to send out intense waves of signals from its learned-routine space which would actually control its external behavior in the waking state. Thus, if the electrical activity in the playal's brain during Event-1, for instance, could be compared with that during its waking state, as with an electroencephalograph (EEG), there would be some resemblance. This resemblance will increase as the capacity of the playal to copy-learn increases, and more and more of what were once waking instinctive functions are performed by culturally-transmitted routines. Thus, in us, the degree of resemblance will be very great, since such a large part of our waking external behavior is driven from copy-learned routines. It might seem "paradoxical," at first, that such a substantial resemblance should exist between the electrical activity of the brain during waking, and that during a part of "sleep," but the resemblance is, evidently, to be expected.

Changing the subject somewhat, it is relatively easy to see that erections of the penis should at least be possible during the Events 0 to W, by noticing that the paralysis involved relates to the blockages of nervous connections to *muscles*. And though males, since time immemorial, might have wished otherwise, the erection of the penis is a tumescence due to an increase in its entrapped volume of blood, which is controlled by a delicate process that relies on the transport of hormones in the blood-stream, rather than on nervous signals, arising in the learned-routine space, that are aimed directly at muscles. Thus, there is no reason, in spite of the generally imposed immobility of the playal sleeper, to suppose that penile erections should not occur during the Events 0 to W. Furthermore, it would seem reasonable that, if erections are to occur during sleep at all, they should occur much more frequently in the part that harbors the much greater intensity of behavioral activity.

By way of summary, it can be said that the features of Events 0 to W match those known to be possessed by REM sleep in striking and significant ways. In particular it can be seen that the Events are such that they:

1: Are always accompanied by profound paralysis of the "motor" parts of the   creature;

2: Should be accompanied by "twitching" in various "motor" parts of the creature;

3: Should be accompanied by signs of emotional upheaval and panic;

4: Could be accompanied by penile erections;

5: Should be accompanied by "waking" activity in the brain;

6: Always begin from the same stage in sleep, Stage 2;

7: Always take place in the same stage of sleep, Stage 1, except for Event-0, which takes place just at the interface of Stages 1 and 2;

8: Are followed by stages of sleep, going from Event-0 to Event-W, which become less and less deep, after starting at Stage 4, the deepest;

9: Increase in duration from Event-0 to Event-W;

10: Have decreasing periods of time between them;

11: Terminate (normal) sleep at the end of Event-W;

12: Have passages between them that resemble NREM sleep.

The degree of agreement between the Events 0 to W, arrived at as I just have, and REM passages, as they have been observed in both young and adult playals, is so complete, that the Events arising in the model must, in my view, be nothing but REM passages. I therefore assume that the model I am proposing provides the beginnings of an evolutionary basis for the origin and development off all sleep, including that in creatures with early brains, and REM sleep in playals.

It is worth noting that one of the effects of REM sleep is to extend the time spent in sleep compared to that spent while awake. Of course, no creature could extend the dangerous period of un-control that is sleep and escape the diligent scrutiny of natural selection without some benefit to show. But it is easy to find the benefit that accompanies the REM extension, since, in much the same way that sleep itself is part of the price paid for the power of a brain, the REM extension of sleep is part of the price paid for the addition to a brain that gives its bearer the power to copy-learn, and to understand, and to increase immensely its capacity to survive.

There are many other important features of REM sleep that have not been discussed. Some of them are well known, and others less so, but all of them would be useful in testing the model, and so could be pursued beneficially if this were a book devoted to sleep alone. But since (I have just reminded myself that) this isn't the case, I have selected the following seven of these many important features for examination:

1. The existence of Event-0;

2. The "periodicity" of REM sleep;

3. The rapid movement of the eyes that accompany REM passages;

4. The sleep of newborn playals, which is said to "begin" with REM sleep;

5. The resemblance of REM sleep to an illness known as "cataplexy";

6. The association between REM sleep and memory;

7. Random departures from the general pattern of the theory described above.

THE EXISTENCE OF EVENT-0

It is of fundamental significance that the present theory leads to the existence of Event-0, and identifies it as a brief REM period linked inseparably to those that follow it. This is so because there is a short, well known EEG feature that occupies just the position identified for Event-0, and which is generally taken to mark the definitive transition to sleep. This EEG feature is usually known as a "sleep spindle," and marks the point beyond which the sleeper would not generally drift back out of sleep. This is, of course, exactly what one would expect of the kind of REM period that the present model associates with Event-0, since that is where the first and maximum burst of blocker occurs that drives the sleeper out of the REM state and into deep sleep.

Such an identification has been made by no other theory of sleep, and there are two reasons for this, associated with features of the theories themselves. The first concerns what is frequently taken by them to be the "periodic" nature of REM passages, to which I shall return shortly. The second has to do with the difference between the shape of the EEG wave that accompanies the sleep spindle, and that which accompanies the later REM periods.

Interestingly, the details of Event-0 and the other Events, provided by the present theory, suggest that rather than being similar, the EEG waveforms should be different, since, in the case of Event-0, the portions of the brain that are relevant to the REM period, and which, presumably, largely determine the characteristics of the EEG waveform, are saturated by the largest burst of blocker, whereas, in subsequent REM periods, the burst of blocker is much reduced, and so the relevant parts of the brain are progressively less saturated. One might therefore think of the mechanism underlying the EEG in the case of Event-0 as essentially non-linear, due to saturation by the blocker, and the mechanism for subsequent REM periods as essentially linear, due to diminishing saturation and ultimately exhaustion of the blocker's effects, with consequent differences in the EEG waveforms.

The identification of Event-0 as a brief REM period is certainly one of the more useful features of the present model, since, as you will see later on, it allows one to understand some otherwise baffling features of sleep, as well as some of the abnormalities associated with it.

You might be interested in the following personal note. The very first version of what is now Figures IV.3a and b did not include Event-0. It was only after I had come to understand the mechanism of the Events 1 to W that I realized there would have to be something like Event-0, just due to the logic of the process. So I put Event-0 where it seemed to belong, and quickly realized that there were EEG and other indications of something like it, just where it wanted to be!

THE "PERIODICITY" OF REM PERIODS

In Figures IV.3a and b (pages 96 and 97), measure the time (distance) between the start of Event-1 and the start of Event-2, and compare it with the time between the start of Event-2 and the start of Event-3. You will find that they are nearly equal. If you compare these two nearly equal times with the time between the start of Event-3 and that of Event-W you will find that these also are nearly equal. Evidently, REM periods give the appearance of recurring at some steady frequency. But although it is clear, from the present model, that this "periodicity" is really just a complicated coincidence, it has frequently been taken as a basic feature of the REM periods, and as a starting point for some theories of REM sleep. This has led to unsuccessful efforts to locate the relevant "oscillator" – the "990-minute oscillator," as it would have to be in us.

Of course, if one begins with the belief that REM periods occur at a fixed interval, then Event-0 will never be seen as such a period, and so the theories that have had REM periodicity as a starting point have locked themselves out of one of the more important observations that one can make about REM sleep.

We can use the present model to help us imagine what the earliest sleep of playals must have been like, for we can see that it must have started in the usual way, then been interrupted with what was probably a very minor Event-0. Not having gone into much deeper sleep, the playal would have drifted quickly back to an Event-W, which would be the only REM activity that would follow Event-0. As playals and their sleep evolved, Event-1 would appear, between the previously occurring Events, and so on to Events-2 and those that follow, but with sleep always ending with an Event-W.

RAPID MOVEMENT OF EYES IN REM PERIODS

Why should the eyes of a playal that are, evidently, positioned by muscles, move during a time that is characterized by extreme muscular paralysis in its motor system? We can see from the model where the answer might begin, since the muscles that position eyes will do so during Event-1 of Figure IV.3a, for instance, so long as the bombardment from the learned-routine space, which lacks the restraining input from proprioceptors, and which commences at time Ta, does not cause the muscles that position eyes to go into paralysis in the same way in which such bombardment causes paralysis in the muscles that drive jaws, and neck, and legs.

If such paralysis does not occur, then, beginning at time T21, when the inner-stream flow comes back ON inside the brain of the playal, eyes will start to move about, since the relation between the muscles that position the eyes and the learned-routine space would then be close to that in the waking state, with signals moving out toward muscles and back from them to the learned-routine space. So the question becomes: Why should the bombardment from the learned-routine space bring on paralysis in the muscles that drive the motor system and not in those that position eyes?

We can begin an answer to this new form of the question by noticing that eyes, even though they can be positioned by signals from the learned-routine space, are not primarily parts of the motor system. Rather, the eyes of playals are very special sense organs (which actually include what are almost extensions of the brain inside them), and, for given velocities and accelerations, they require *fixed* applications of forces to change their positions. Thus, to the extent that they change position at all, they require muscular activity of a type that could be largely defined and programmed, as must be the case in reptiles, with the programs remaining the same from one generation to the next.

From this we can see that eyes, precisely because they are vital as sensors for executing most forms of external behavior, will, already at the stage of the reptile, have almost all of their positioning routines determined genetically, and be programmed in this way. Thus, the extent to which the learned-routine space, when it emerges, need become involved in the actual control of the muscles that position eyes, is limited to relatively minor modulations of the basic genetic program.

By contrast, we can notice that the emergence of the learned-routine space is associated with new, copy-learned external behavior, which relates primarily and directly to new routines for the movement of jaws, and neck, and legs, and torso, for instance. Such behavior must be expressed through the highly variable forces that need to be exerted by the muscles in these parts, and this leads to a critical dependence on the signals coming back from their proprioceptors to the learned-routine space, in an inner stream. Consequently, in the absence of such signals coming back from jaws, and neck, and legs, and torso, much more than in the absence of signals from muscles for positioning eyes, the learned-routine space will have a tendency to send out exaggerated bursts of signals which ultimately overwhelm the capacity of the muscles concerned to respond.

In view of all this, its seems reasonable to suppose that the tendency of the learned-routine space to overwhelm the muscles that position eyes within the fixed, internal environment of the head, because of a missing inner stream, must be far less than is the case with the muscles that drive highly variable force-generating motor functions, which actually give expression to new, evolving behavioral routines for the playal, as it builds its store of culturally-acquired routines, which must run in an external environment of extremely variable demands.

Looked at a little differently, it seems that copy-learning added relatively little to the ways in which eyeballs are positioned, but it added a very great deal to the behavior of those parts of the playal that actually determine external behavioral performance. In effect, there is relatively little need for signal traffic between the muscles that position eyeballs, and the learned-routine space; almost everything needed to position them was in place before the learned-routine space came on the scene. Thus, the limiting flow of signals from the learned-routine space into the muscles that control the position of eyes is far less than that into the muscles that sustain motor members of the body against huge, variable, external forces.

It is, in my view, this difference in the ultimate rate of flow which leads to the saturation and paralysis of muscles in the one case, and the continued ability to follow in the other. And so, in the movement of eyes in REM sleep, we encounter the paradoxical outcome of the loose coupling between the learned-routine space and the muscles that position eyes, since, in the fact that eyes actually move, rather than being as still as limbs, we see a reflection of the relative looseness of the coupling between the driver and the driven. It is only when we go below the surface, and look at the underlying workings of the model, that we can come to see how it might be that much coupling to limbs can lead to little movement, while little coupling to eyes can lead to much movement, and so resolve the paradox of REM sleep, when the movement of small eyes contrasts so sharply with the stillness of large limbs, which, because of their great flaccid size, tend to define the REM sleep of playals, for the casual watcher, as a "quiet time."

THE SLEEP OF THE NEWBORN

There are a number of references in the literature on sleep to pieces of research which are taken as showing that "sleep of the newborn begins with a REM period." If one thinks about this statement a little, difficulties begin to appear, for reasons that I shall now explain. Let us start with the assumption that waking is one state of a creature, and sleep another identifiably different state. If this is not assumed to be so, then the "begins" in the statement, "sleep of the newborn begins with a REM period," can refer to nothing, and the whole statement is not worth making, since there would be no identifiably different state following waking, namely sleep, the *beginning* of which one could identify.

Now it is clearly a fact, which the simplest of observation can confirm, that creatures known to sleep, being massive biological agglomerations, cannot change from one identifiable state to another identifiably different state in zero time, that is, cannot change from one identifiable state to another identifiable state without passing through a sequence of transitional states which allow the various parts of the creature to assume their final, identifiable states.

What this implies is that, in constructing a description of the behavior of a biological agglomeration, that is, in trying to understand it, we will always have an incomplete description, which is to say, fail to understand it at the level of the description, if we confine the description to just two states. As a minimum, we have to keep track of *three* states: one identifiable state, another identifiable state, and a transition state that connects the two. Unless we accept this constraint on the description of biological agglomerations, that is, carry along an explicit transition state between any two identifiably different states, we invariably end up "showing" that the two clearly different states are identical, since only identical states can be separated by no transition state.

Clearly, one of the factors that has allowed a statement of the type just discussed to gain as much uncontested prominence as it has, is the absence of an evolutionary model which not only prompts one to suspect the existence of the quite fundamental difficulties inherent in the statement, but which also, and, even more usefully, allows one to begin to surmise what might be the nature of the finer structure of the continuous transition that always must convey a creature from an identifiable state of waking into an identifiably different one of sleep. I shall demonstrate such a surmise as to the nature of the finer structure in the discussion of the illness "cataplexy" that follows, and then suggest an extension to the newborn.

REM PERIODS AND CATAPLEXY

Rather than begin with a definition of cataplexy, let us refer to Figure IV.3a on page 96, and imagine that we have come on a playal that is not normal, because of what I shall call a "leaky" producer of sleep-inducing blocker in its brain. By this I mean that, at some point in its life, the playal's brain begins to produce small quantities of blocker at even the lowest levels of stress, and that, as the playal ages, its brain produces higher and higher levels of blocker in this way. As we can see from Figure IV.3a, this would mean that, in what could be taken to be the waking state of the playal, it would have begun a long transition to sleep, in which it would remain continually just below the lower reaches of Stage 1, but would not go into a state of sleep because the effect of the blocker would not, except at the usual stress level for the induction of sleep, be able to induce it. Indeed, if the increase in "leakyness" is sufficiently gradual, taking many months or even years to develop, the playal might appear "normal," even though, to maintain such "normalcy," its sensitivity to the blocker is decreasing, that is, the lower level of its Stage 1 is rising toward the lower level of its Stage 2, so that its "waking" state is moving closer to the level of blocker that can induce Event-0.

If the "leaky" production of blocker continues to increase, the "normal" playal will eventually reach the situation in which its Stage 1 will have contracted so that the "waking" level of the blocker in its brain is marginally close to the lower limit of Stage 2, that is, the level corresponding to time T12, which marks the onset of Event-0. When it reaches this state, even the smallest increase in blocker production, brought on by some chance occurrence of minor stress, will set off the REM period associated with Event-0. If, now, because of the ongoing leaky production of blocker, the capacity to produce a large burst is limited, this REM period might last longer than it normally would, since it will take longer for the signal-flow from the learned-routine space to be turned off at Tz.

Thus, what is normally a relatively brief episode, at Event-0, which goes practically unnoticed, can now become a REM period of long duration, including an extended period of paralysis, and all the other features that accompany such an event. Evidently, in such a case, we would appear to have a situation in which "the sleep of the playal begins with a REM period." But this way of stating what might be observed, just in the immediate context of the REM period alone, would miss the crucial fact that, for this to appear to happen, the entire blocker production process in the playal had to be so far from normal that the creature must have lived in a continuing state of transition to sleep ever since leaky blocker production began. With such an extended transition to sleep, which might have lasted months or even years, any single passage to a REM period is not usefully viewed as an event with which sleep "begins." Furthermore, one can see how a readiness to accept the view that sleep can actually "begin" in this way would hinder the search for the basis of the related illness.

Such an illness, the symptoms of which include falling asleep, perhaps briefly, with a passage through Event-0, followed by some version of Event-W, including paralysis of motor parts of the body, and all of which begins suddenly after the playal chances upon some stressful incident while "awake," is known as "cataplexy." The general context of some sort of waking drowsiness in which cataplexy would seem to have to be situated is known as "narcolepsy."

We can see from the foregoing that cataplectic events should display all the characteristics of periods of REM sleep since, according to the model, this is precisely what they are. A large quantity of careful measurement has shown this similarity of characteristics to exist, although, without a model of sleep, the resemblance between REM sleep and cataplexy has tended to remain something of a mystery.

We can also see, from the model, that cataplexy should be found in all species of playals, and could be expected to have a hereditary component, since some of the departures from the normal could be due to mutations that affect the functioning of the relevant parts of the model, such as the producer of the blocker. But although the necessary brain-structure for supporting cataplexy would be present in all playals, it would seem unlikely to be evident in tree-climbing primates, or cats, for instance, since repeated uncontrolled attacks of sleep would lead to falls, and almost certain death, thus removing from the hereditary process those potential parents that would perpetuate the illness. It would be similar in all but flightless birds. On the other hand, cataplexy would seem likely to be present in essentially ground-living domesticated playals such as dogs, because they are exposed neither to the risks of lethal falls nor to the attacks of predators that would be likely to profit from the random periods of helplessness that characterize cataplexy. As it happens, hereditary cataplexy has been identified in dogs, and it is interesting that the model of sleep that is being advanced should be able to suggest the existence of such an illness in them.

One can also see, from the nature of what is basically a chemical model, that the symptoms should be sensitive to the introduction of drugs which would affect, for example, the production of the blocker. But it is also clear that attempts to alter the nature of cataplexy, by means of drugs, will almost certainly alter sleep at other times as well, and will make a successful, isolated assault on just cataplexy, as the serious illness that it can be, difficult to realize.

Returning now to the sleep of the newborn discussed in the previous section, we can see that the model allows for a number of possible continuous transitions from waking to sleeping which would avoid the difficulty of appearing to believe that the sleep of the newborn could "begin" with a REM period. For instance, it could be that one more of the many incompletenesses already known to be associated with the brain of the newborn, related simply to its naturally incomplete development, might be the presence of a "leaky" producer of blocker, which holds its brain in about the same state as that of an adult suffering from narcolepsy. Thus, until the maturing of the playal's brain gradually fills in this incompleteness, as it eventually fills in many others, the frequent sleeps of the newborn would simply be cataplectic sleeps, which would appear to begin with a REM period. As the filling-in takes place, the sleep of the newborn will assume the form of that of the infant playal, aged about three months or so, whose sleep is well known to conform to the adult pattern.

But it is interesting to wonder whether such a narcoleptic state immediately after birth would not simply be an extension of a state of narcolepsy which must exist during the development of the mammal fetus generally, and which finally smothers the trauma of being born in the daze of cataplexy. This would have formed a reasonable, fundamental part of the general evolution of mammals with ever-enlarging brains and skulls for holding their beginnings at birth, and so would find its clearest expression in the newborn human.

REM SLEEP AND MEMORY

The literature on sleep bears evidence of a great deal of research on a possible link between REM sleep and memory. Some of this work reflects the belief that perhaps REM sleep might even exist for the purpose of realizing memory. Although not all the work is set in this context, much of it carries, if only tacitly, a seeming belief in the "purpose" of REM sleep which would tend to be unacceptable in any reasonable evolutionary argument, and which the model presented here would certainly lead one to question.

However, it happens that the present model *does* suggest a link between REM sleep and memory, but the way in which this comes about can be placed well outside the kind of purposive context just mentioned. Unfortunately, anything approaching a full treatment of this link would take far more space than is appropriate here, so I shall place my emphasis on setting up the context in which the link is situated, and then, as promised earlier, sketch the nature of the evolutionary link itself between REM sleep and memory, and how it might have arisen.

Although it might be difficult to arrive at a fully satisfactory definition of "memory," we can start by stating that "memory," the phenomenon, is an expression of the influence of past events on present behavior. Associated with the phenomenon "memory" is "a memory," the thing, which is taken to be the physical basis of "memory" the phenomenon, and as the place in which an unexpressed instance of the phenomenon, "a memory," can reside. I shall use "memory" in all of these senses, relying on the context as a guide to its meaning.

We can see immediately from this that it will not be useful, in an evolutionary explanation, to imply that creatures are evolving "so as to acquire memory," or that the "purpose" of REM sleep is to yield or to expand memory, since evolution is driven by immediate behavior and its relation to immediate or past behavior and environment, rather than by behavior or environment that is yet to come.

Consequently, if the present behavior of a creature should begin to express the influence of past events, this will have to be viewed as arising in some underlying feature of its biology that is not in any way committed to serving as a basis of memory as such, but becomes such a basis purely by virtue of the intrinsic nature of the underlying feature itself, coupled to the evolutionary circumstances of the creature, and the way in which natural selection can filter the line of creatures involved, from one generation to the next.

Such a feature of the underlying biology of every creature is to be found in the fact that even though chemical reactions can go backwards and undo themselves, that is, are reversible, they take some time to reverse; thus, for memory to emerge in the behavior of some creature, the persistence associated with this completely unavoidable delay would have to lead to some survival benefit on which natural selection can operate. The emergence of memory will then depend on three factors: the first is the completely unavoidable and variable persistence of chemical states; the second is the possible benefit of this persistence in the survival of some line of creatures; and the third is the filtering effect of natural selection on this possibility, generation after generation, as mutations provide the basis of more forms of chemical persistence.

Having arrived at such a conclusion, it is clear that the most basic form of biological memory is to be found in the genetic process itself, by which both the form and the behavior of elementary creatures are determined. The unavoidable chemical persistence is found in the tendency toward an unchanging structure evident in the genes themselves; the influence on, and the possibility of beneficial behavior comes from the fact that genes largely determine behavior; and the filtering by natural selection is evidenced in either the continuance or the extinction of any particular line of creatures, as the ongoing stream of mutations in their genes contributes to the one outcome or the other.

Evidently, the net effect of all of this is that the events in the lives of all the antecedents of a particular creature influence its present behavior, as is required by the notion of memory, stated earlier. We can see from this that, in a very fundamental way, *all* biological systems are endowed with a kind of memory that preceded the appearance of REM sleep.

It is important to notice that it is legitimate to include, in what I am calling "behavior," the *development* of each creature starting from the single cell that serves to define its beginning, since, on the scale of the cells themselves, which make up the agglomeration that defines a creature at any time, their individual development constitutes very real "behavior," albeit behavior devoted to development rather than to the external performance of the creature as a whole. This serves to highlight the artificial nature of the distinction that tends to be made between development and behavior.

When viewed in this way, we can see that the memory associated with the necessarily continuous development of a creature, which must involve the continued influencing of the behavior of some thousands of billions of cells in a creature of about our size, must considerably exceed any memory that could ever be associated with the execution of external performance during a lifetime. In effect, if the memory associated with genes has the capacity to control behavior as development, it also (easily) has the capacity to control behavior as external performance, which, as I have just indicated, is an artificial distinction in any case. Thus, it is not difficult to see that the present behavior of any creature, in the full sense of both its development and its external performance, could be overwhelmingly controlled by the influence of genes, that is, by *memory* related to events that occurred in the lives of its antecedents.

We can therefore reasonably imagine a creature, which I shall take to be a reptile, whose behavior is completely controlled by its genes, that is, by the influence exerted on its present behavior by the events in the lives of its antecedents, with only the most minor variations as these would express any mutations that appeared in it, but which would themselves be expressed as genetic control. Such a creature would have no learned-routine space, since it would have no capacity for copy-learning. Indeed, to the extent that it is under total genetic control, it has no need for any behavior-determining memory other than that expressing its genes—there is a prescription for behavior in every possible circumstance in which the creature can find itself, as the events in the lives of its antecedents, filtered through the extinction and survival of natural selection, have defined "possible circumstance."

If we now compare a playal with this creature, then the ability to effect copy-learning requires a way in which events occurring at various times in the playal's *own* life can influence its *own* subsequent behavior. This clearly requires that the behavior of the playal express the emergence of some new way in which past events can influence its present behavior, and this must require the emergence in it of a new kind of memory, different from the kind that is locked directly into genes.

Evidently, this new memory, which must accompany copy-learning, as discussed in the preceding chapter, is the one in which the inner-stream flow is stored, and is the basis of the functioning of the learned-routine space. Since, according to the model of sleep in Figure IV.3, REM sleep is a development attached directly to the emergence and existence of the learned-routine space, it would be surprising if there were no connection between the development of memory associated with the inner-stream store, and the occurrence of REM-sleep episodes.

And, indeed, already, at this point, we can make a statement about the possible relation between REM sleep and memory, for we can see that, if there is to be any relation between them at all, it will have to be one between REM sleep and the kind of memory that can support behavior that is associated with copy-learning, rather than with direct genetic control. This, as it happens, is one of the more significant findings of the work to which I referred earlier, and, in that context, the finding is often stated as being that the link between

REM sleep and memory seems confined to those areas of behavior in which the creature is "unprepared."

We have now assembled enough by way of context for me to proceed with the sketch promised earlier of how the connection might have arisen between REM sleep and memory. To begin, we need to remember that the inner-stream memory will have to be reset from time to time because the configuration information stored in it has ceased to be current, and that this is the basic underlying condition that induces sleep. However, for the reasons given before, the complete reversal of the state of the chemical entities that constitute the memory will take time to occur, and, in principle, will never be quite complete. Thus, there are basically two ways in which a creature that sleeps, that is, a creature with a brain, can evolve: it must either write over what is left in the memory after a period of quiescence, or it must find a way to add capacity to the memory ahead of the incompletely faded traces in the memory.

In the case of the reptile, there is no benefit in anything but writing over the fading traces, since it only needs to refresh its configuration store, and get back to using its genetic memory to survive. It has no use for a permanent store of the proprioceptor information flowing in its inner-stream, because nothing in the past of reptiles has led them to acquire the capacity for copy-learning; that is, they are not faced with the facts of primal play.

But the gradual emergence of the playal would have completely changed the significance of simple overwriting, since, with the first glimmer of the possibility of copy-learning emerging in play, the fading trace in the inner-stream memory becomes precisely what the playal needs to realize the entire new set of possibilities that flow from being able to retain and repeat the behavior of which the only usable nervous record resides in its inner-stream memory.

So we can now begin to see how the benefits could arise that would sustain a slow shift, from the previous overwriting of the inner-stream memory combined with genetic control in the reptile, to the retention of the contents of the memory and its expansion, now combined with copy-learning in the playal. But one must still ask how such a change could begin as a reasonable evolutionary process. I believe that it began when a mutation led to an increase in the retention-time of the inner-stream memory in the very earliest playals, thus making overwriting increasingly impractical on the one hand, but retention of behavioral routines more practical on the other.

However, this, by itself, would have led the emerging playal lines to certain extinction, for when the inner stream begins to be retained permanently in memory, there would have been no way to reset the increasingly inaccurate version of its configuration that is available in its brain, and hence no way to support a brain. Thus, this development, by itself, constituted no favor to playals, the "purpose" of which might have been to "help them to develop" copy-learning; rather it represented a growing threat of extinction which, without some *other* development, would have led to their disappearance. What saved them, then, was the only thing that could, and that was the laying down, during each period of REM sleep, of new "layers" suitable for use as inner-stream memory.

So, what would stimulate the growth of new layers of memory-space during REM sleep? Probably the extreme conditions of stress and blocker production that accompany REM sleep today, and that must have accompanied it ever since the first time that a twitch of it disturbed the quiet sleep of a creature then departing the ranks of reptiles.

This would then resolve a big paradox and solve a big mystery, all at once. The big paradox, which I have highlighted earlier, relates to how a playal brain can have the ability to start with a clean configuration memory that contains no record of the past inner stream, after each sleep, and so continue to make a brain possible, while, at the same time, can have the ability to retain the record of the past inner-stream episodes indefinitely, and so begin to make copy-learning possible. The mystery has to do with the retention of the order of past events, since, what the progressive extension by layering of memory-space does is allow the relative dating of the contents of memory according to the relative places of the layers in which they are found, much as geologists would learn to do relative dating of past events millions of years later.

By way of summary of this brief sketch, then, what the present model says is that the link between REM sleep and memory is, first of all, located in the area of memory associated with copy-learning only, which is to say, with understanding. This is the kind of learning that we also think of as "socially transmitted," or, as is sometimes said in the literature on sleep research, learning for which the playal is "unprepared."

Further, the link takes the form of some kind of chemical "formatting," during REM sleep, of successive layers of brain-space. This renders each successive layer suitable for recording the *new* inner-stream flows that occur between one sleep and the next, and "fixes" them permanently, layer by layer. This layered recording and storing make possible both the permanence necessary for copy-learning, and the transience necessary for configuration updating of a brain, all in the same brain. It also allows sequential dating of events. Thus, REM sleep is not linked to the processing of events into memory, this goes on while the playal is awake; rather, REM sleep is linked both to the layered formatting of brain-space, which makes new "layers" of processing possible, and to the "fixing" of each previous new layer of brain-space into permanent memory.

It would seem unlikely that the amount of brain-space formatted during any single sleep would be critically connected to the space required between sleeps, but, if the process of formatting should cease, the playal would soon begin to run out of inner-stream memory, thus inducing sleep and the appearance of being unable to repeat "recently copy-learned behavior." However, this would more properly be seen as a growing inability to store copy-learning altogether, in a memory that no longer has room in which to hold the inner stream that it should record.

We can usefully pursue the structure of memory arrived at in this way a little further, and see what it might suggest regarding our ability to recall dreams. What we have to account for is the common experience, supported by careful observation, that we recall only those dreams that occur in the passage of sleep that directly precedes waking.

Using the structure just described, we can imagine two adjacent layers of memory that carry the streams associated with the copy-learning occurring before and after a period of sleep. In such an arrangement, the "experience" associated with dreams occurring during the period of sleep would have to be stored in the interface-space between the two layers of memory. Given this situation, it is surprising that we can recall any dreams at all, since they would have to be stored in a transitional memory space that would seem likely to lack the "formatting" necessary in a genuine layer of memory.

What would seem to be the case, then, is that only the dream sequence that is adjacent to the edge of the period of sleep that precedes waking is located in a memory space that is sufficiently well formatted to support memory, and allow recall. We can achieve this effect in the following way: assume that (a), formatting is associated with the early part of each REM period, otherwise sleepers awakened in REM periods would not recall the adjacent dreams as reliably as they do; and (b), that the activity in each REM period is such as to overwrite whatever was stored in any previous part of sleep in the transitional memory space, otherwise all the dream sequences in sleep would be recalled with the same clarity as the one that precedes waking. With these assumptions, it would also be the case that sleepers awakened in NREM passages would recall dreams, as indeed they do, although, soon after the discovery of the REM phenomenon, it was reported that dreaming occurred only during REM periods.

What is interesting about all this is that the model of the relation of memory to playal sleep that has been proposed allows a relatively simple way of beginning to account for one of the most baffling aspects of sleep, that is, the highly selective way in which we seem to have been allowed to recall the dreams that accompany it.

Random Departures from the Model

Some of what I am about to say will seem immodest, but, if you were to search the literature on sleep, and compare what you find there with what I have been saying about it, you would conclude, after a while, that the model I have been advancing allows one to account for much of what is known about sleep's general features. Indeed, you would notice, for instance, that the model of sleep summarized in Figure IV.3 is the first to bring together an evolutionary theory with any sort of consistent process relating to the emergence of REM sleep, so as not only to explain much of what is known about it, but also to show its evolutionary connections.

But what you would also notice is that, although the agreement between the model and the general nature of sleep is good, the actual data gathered from sleepers show large numbers of excursions that depart from the nice, smooth picture suggested by the model. Does this have to be so, and, if so, why?

I think it has to be so, and that this relates to the enormous complexity of even the "simplest" creatures that sleep, and the fact that they never "just sleep," since sleep must allow of being disturbed, so sleepers can be awakened for purely situational reasons, which can arise either inside or outside them. Hence, the complexity of the inside of the sleeper, combined with the semi-openness of even the deepest sleep, will always leave room for bursts of this and bursts of that, which disturb the nice smooth lines of a purely theoretical sleep but which, presumably, also suit the real live sleeper.

Stated differently, one might say that an attempt to produce a model of sleep that could account for all the little details that appear in the records of real sleep would have to become a model of something else, which would have to be, all at once, a "something" more embracing and more detailed than sleep. This does not mean that either the present model, or some other, will be a final statement, without the possibility of expansion downward into more detail. It does suggest, though, that what one will be able to include in a model of sleep will never be able to explain all the detailed aspects of the data that will be collected in the name of "measurements on sleep." There will always be bursts of this, and bursts of that, which fall outside the explanatory power of a model of sleep, because they can only be explained by a model of something else.

Notwithstanding all of this, the explanatory power of the model being advanced here is quite considerable, and so it is important to look in some detail at four assumptions that are peculiar to it, namely:

1. Stress-creating errors accumulate inevitably in an inner-stream store;

2. The brain produces a (blocking) substance that serves to induce sleep;

3. The playal brain is the basis of REM sleep;

4. The inner stream and its store are real.

STRESS-CREATING ERRORS ACCUMULATE

The assumption to be discussed here flows from what I believe to be really a self-evident prior assumption, according to which any brain, which controls a portion of the external behavior of some creature, will have to store a current representation of the configuration of its host. We can go further and say that, not only must there be such a configuration store, but, even in a highly evolved brain such as ours, the initial part, at least, of such a store must still be located somewhere in its most primitive part (brainstem), since this absolutely essential store must have evolved with the earliest brains, as integral parts of them. So what I wish to discuss in some detail here is the assumption that errors must inevitably accumulate in this configuration store, creating stress as defined in Chapter I, and inducing sleep.

There are at least four ways in which errors can find their way into the configuration store. The first stems from the fact that growth is continually occurring in every creature. So its configuration is continually changing, because the parts that define it are continually changing. Thus, even if the representation at any instant could be free of errors, they would soon begin to creep in, since what is represented is changing. The second way for errors to originate is in the nervous circuits that carry the configuration information to the brain. The third source of errors comes from the fact that the creature changes configuration due to external forces of various kinds. The fourth source is located in the store itself, since, due to the biological activity that must impinge on the store, it will experience disturbances in what it has already stored. Such errors are unavoidable, and will eventually render a brain not just ineffective, but destructive, if some process for removing them is not available. It is, as you have now seen, this process for removing errors that can form the basis for a comprehensive theory of sleep.

The errors arising in growth are particularly interesting to focus on, since these would accumulate at a very much higher rate in very young creatures. In this way we come on what must be a partial clue to the high frequency of sleep in the young, and this clue deserves a more complete pursuit than I intend to give it here.

A Sleep-Inducing Substance

It is convenient, given the extent to which the model has now been developed, to begin by asking what a sleep-inducing substance must do, since, even with a quite detailed model of sleep, the question is still not easy to answer. But the necessity of having a model of sleep in order to answer the question can be seen by noticing that, in the absence of one, it might seem, for example, that there would be a need for *two* kinds of substance, one to induce the "early phases" of sleep, and another to induce REM sleep. However, one can see from the present model that sleep should require only *one* substance to induce all its phases, since they can be accounted for by appealing to the *order* in which certain nervous blocking and unblocking events (must) occur.

Equally important, I believe, is the fact that the model allows us to conclude that REM periods are much more a manifestation of how parts of the brain interact when interconnected in a certain way than of the effect of a substance on the parts themselves. A useful metaphor might be: REM periods are more a manifestation of what happens when the switches in an exchange come to be arranged in a certain way, than of what happens if we drug the subscribers. This leads to a conclusion that might seem overly fastidious, but is important, and which is as follows: while it should be possible to induce a REM period artificially, it is far from clear either what an "artificial REM period" might be, or how one might induce it.

Such a conclusion is, I believe, especially relevant to the section of J. Allan Hobson's recent book *SLEEP* entitled "The Experimental Control Of Sleep," in which he refers (page 137) to inducing "artificial REM periods" with injected substances. What I find unclear is whether the interventions of the experiments, as he sees them, are addressed at the exchange, the subscriber, or both. It should be added that, although I shall not show it here, the results described by Hobson can be placed well within the context of the ("exchange") theory presented here.

Looked at more generally, there is nothing about the production of a sleep-inducing substance in a brain that would seem the least unlikely, as knowledge of brains stands today. Furthermore, the whole time-scale of sleep, and the clear picture that the later REM periods present, of harboring an essentially orgastic process that is basically running down, suggest that, fundamentally, sleep in the playal is controlled by an exhaustible source of blocking substance, which the new complexity of the playal brain does, in fact, succeed in driving to exhaustion, when it sleeps, and before it wakes.

It is important to notice that, according to the present model, the state in which sleep is situated at any time is determined almost entirely by the state of decay of the blocker and its related effect. This provides a basis for understanding the "momentum" that sleep tends to display, in that brief excursions to waking are common, without the total disruption of sleep. This could only happen if the brain can "remember" the state of sleep in which it was before awakening. Evidently, the necessary "memory" is provided by the slow decay of the blocker and its effect, allowing sleep to be resumed where it left off, if the period of awakening is not so long as to allow substantial decay of the blocker, with the consequent "forgetting."

This is a good place at which to re-emphasize the importance of being able to reduce the model to one requiring just a single substance for inducing all the phases of sleep. In the first place, it provides a certain kind of assurance that one has come on the correct form of the model, or, at least, that it would not be easy to find a simpler one based on sleep-inducing substances. In the second place, the fact that the single substance seems to be associated with the general development and response to stress has considerable explanatory power, since it allows certain common-sense feelings about sleep to be placed in a consistent framework that can also accommodate its more arcane aspects, which have only been revealed as the result of careful measurement and observation within the last fifty or so years.

Further, this link, between stress and sleep, allows us to pin down another connection, between the comic and the tragic, which tends to float in common sense, but still lacks confirmation via more rigorous examination. The nature of this connection begins to emerge from the fact that, strange as it might seem at first, cataplectic sleep is preceded most frequently, not by obvious stress or rage, but, by, of all things, *laughter.*

So what the link provided by the model begins to tell us is that laughter could be nothing more than a manifestation of a condition of *ongoing stress*, and that, in cataplexy, we have the most sensitive indicator of this that is available to elementary observation. This extreme sensitivity is due simply to the incipient state of sleep in which the illness suspends its victim, and the resulting chronic proximity to the normally very minor REM-sleep outburst that is associated with Event-0, and which can therefore be set off by minimal increments in stress.

It also tells us what to look for as the evolutionary precursor to laughter, as it appears in us. This would, evidently, not be a concomitant of "amusement," but rather of fear, frustration, and stress. When viewed in this way we can begin to understand the "strange" resemblance between the facial expressions that accompany laughter in us, and those that accompany fear and rage in many of the playals we know well, such as cats, dogs, and certain primates.

This allows us now to begin to see why laughter accompanies tickling, and confirms the assertion, made in Chapter II, that tickling is simply another attenuated form of killing activity, which forms part of the general repertoire of killing behavior that is the basis of primal play.

Further, it serves to explain the essential two-part structure of *jokes*, which must promote laughter because of the stress-generating "lead-in," and for which the "punch-line" that comes at the end serves to emphasize the stressful nature of the "lead-in" that preceded it. It serves to clarify the link between riddles and jokes, since it shows the mystery of the riddle, which comes at its beginning, as the essential stress-generator that it must be, in order to induce the laughter that follows the "answer," in a really good riddle.

Indeed, what we are allowed to see, in the link that the model establishes, between a single, stress-related substance that induces normal sleep, and cataplectic sleep that is most often preceded by laughter, is the basis of the haunting feeling that, somehow, all comedy is nothing but stressful tragedy in disguise.

And the model also allows us to see that what the sleep-substance did was far simpler in the sleep of reptiles, before the playal arrived with its copy-learning, and its understanding, lodged in a learned-routine space. Indeed, it would seem that the brain of the playal stumbled on sleep, and *almost* fell; and there might even be playals among us today who, in the early hours of some quiet morning, will stumble over their missing bodies, and, while held in the panicky manacles of sleep…fall.

PLAYAL BRAIN AS BASIS OF REM SLEEP

As you might recall, at the beginning of the discussion of REM sleep, I, seemingly quite arbitrarily, limited the discussion to sleep in playals alone. However, you will have seen, in the foregoing discussion, the crucial role that the learned-routine space (which is a concomitant of copy-learning, and which is, in turn, a development flowing out of play), occupies in the way in which the model goes from the sleep of reptiles, which does not include REM sleep, to the sleep of later creatures which does. So you can now appreciate the importance of the limitation to playals, and the reason for it. Indeed, given the power of the model that has been assembled within the framework of this limitation, one might even be tempted to say that the limitation has been shown to be justified, and, "therefore," REM sleep must be a manifestation of the playal brain exclusively.

But even though I intend to show that REM sleep is such a manifestation, the conclusion cannot be drawn with assurance quite so easily, since, even though there is always clear evidence of REM sleep in creatures that clearly meet the criteria for playals, some REM sleep is also found in creatures that appear to be outside the group that meet the criteria. So what I shall do now is review this area of the subject generally, and show how what might appear, at first, to be exceptions to the notion that REM sleep is an exclusive manifestation of the playal brain, can be accounted for.

It would be useful to recall that the criteria that creatures must meet to be playals are that they must be animal-eating animals which rear their young in close proximity to themselves. The first tendency might then be to say that playals are simply a subset of mammals. But, as I showed in Chapter II, this would neglect the whole class of birds, and a very significant neglect it would be, since we can find among them many that are animal-eaters and whose young are, evidently, reared in close proximity to parents.

Thus, we should expect that there would be playals among birds as well, and not just among mammals. In this way we come on a substantial part of the explanation of what might seem, at first, to be a massive exception, because REM sleep is not found in mammals alone, but in birds as well. Indeed, in this way, the existence of brief periods of REM sleep in birds, which might seem, at first, to be an almost insurmountable exception to the playal basis of the present model of REM sleep, turns out to be a strongly supportive feature of it. But, in view of the paralysis that is necessarily associated with REM sleep, we can see that there would be a severe limit on the *duration* which such sleep could ever attain in surviving species of birds, since, except for those late-developing species that have reverted to flightlessness, paralysis would lead to lethal falls, and limit the extent to which such development could evolve. This would place a fairly definite limit on the extent to which the playal brain in birds could develop, and doubtless accounts, at least to some extent, for the extended developments in the playal brain that proceeded in mammals, and not in birds.

We now come to a more subtle "exception," which has to do with the fact that some mammals that are not animal-eaters, do show evidence of REM sleep. In a general way, one can say, however, that the extent of REM-sleep activity tends to be roughly correlated with the extent to which the mammal can be observed as being predatory. Here, the resolution of the exception must turn on the evolutionary origin of mammals, and of REM sleep.

So far as is known, all mammals evolved from a line of creatures that emerged from reptiles, and were eaters of insects. It is here that the "exception" begins to be resolved, because such an evolutionary origin of mammals would make of them, if not all *presently* eaters of animals, then, certainly, the *descendants* of eaters of animals. It therefore seems reasonable that, even in mammals which, in their present evolutionary state, do not meet the criteria of "true" playals, because they are not currently predators, one might expect to find traces of a distant insectivore ancestry, which, at the time of this distant predatory behavior, developed all the features of the playal brain, beginning with primal play, and proceeding to copy-learning and the associated learned-routine space with its accompanying REM sleep, in exactly the way that the present model of the evolutionary development of sleep would prescribe.

As some of these creatures deviated from insectivore to herbivore behavior, they carried, in the genes that determine the structure of their brains, hereditary evidence of the stage that copy-learning had attained when the deviation began. In the case of those mammals that deviated more and more toward predation and the capture of larger and larger creatures, the benefits of copy-learning made possible the evolutionary development of an increasingly elaborate learned-routine space capable of supporting increasingly complex demonstrations of copy-learning, selfness and understanding, and hence clearer and more vivid episodes of REM sleep. It is interesting to notice, though, that once the learned-routine space is established, its development could continue even in the absence of ongoing predation, although this would hardly be expected to be the rule.

Thus, in this quite reasonable and important evolutionary context, all mammals can be regarded as playals, but some are, as we can observe them now, only vestigial playals, and this is reflected in the highly variable forms of REM sleep that they display, even though predation, and play itself, might long ago have ceased to have any significance in the lives of the parents and young of their herbivore antecedents, and have completely disappeared. These vestigial playals have frequently become the prey of those other playals that are active predators today, and it is those active predators that are easily seen as meeting the criteria of "true" playals, not only in the readily observable episodes of play between parents and young, but also in their consistently vivid episodes of REM sleep, which reflect the highly developed capacity for copy-learning that the demands and the benefits of active, selective hunting and killing have continued to engender in them.

Seen in this way, no exception or mystery remains in the fact that, for example, the herbivorous guinea pig displays REM sleep, because its brain must contain vestiges of the playal brain of its ancestors. Further, given the intrinsic benefits of copy-learning, the development of the learned-routine space of the guinea pig's brain might have continued even after the predatory requirements for its emergence had ceased to exist. And so the playal components of its brain would be vestigial purely in the sense that the predatory basis for their emergence, rather than the beneficial basis for their continued development, had disappeared. Evidently, the beneficial basis for the continued development of the playal components of its brain might not be as substantial for the guinea pig as it is for a hunter and killer, but it is not entirely negligible either. Thus, we might find, in the brain of the herbivorous guinea pig, a well-developed capacity for copy-learning which even surpasses that of its predatory ancestor, accompanied by well-defined periods of REM sleep, which are the inescapable concomitants of such a capacity.

Related "exceptions" are those birds that, today, are not predators, but that show traces of copy-learning and REM sleep. Here, the situation must be exactly as is that with the derivatives that formed the ancestors of birds were certainly predators. Thus, we can say that all mammals and all birds can, in an important evolutionary context, be regarded as playals, but of which some are purely vestigial, their antecedents having deviated, long ago, from the predation that would make them recognizable as "true" playals, today.

All that remains, now, is an interesting nuance in the matter of "exceptions," since, although I am making the case that the paradoxical combination of predation and the rearing of young were absolutely necessary for the emergence of play, followed by copy-learning and REM sleep, there is no reason to expect that every evolutionary excursion that involved the existence of predation, accompanied by the rearing of young, would necessarily have led as far as copy-learning and REM sleep, even if the paradox had been resolved by the emergence of primal play as an element of behavior.

It would therefore add another "exception" to the list of those already resolved, and be reassuring, if we could find a creature that would, ideally, be situated, by virtue of its state of evolution, somewhere between reptiles and mammals, and that would be an insectivore, rear its young, and still not show signs of REM sleep. As it happens, the "echidna" is just such an animal, being a member of the post-reptilian but pre-mammalian family known as "monotremes." Thus, the echidna, with its lack of REM sleep, when viewed in this way, emerges not as an exception, but simply as evidence of the small steps in which evolution must proceed, since, in it, we can see, resolved, the step between the existence of primal play, which accounts for the fact that this young-rearing predator exists at all, and the emergence of copy-learning with REM sleep, which is primal play's natural sequel, and with respect to which the echidna just falls short.

With this, one can see that there are really no remaining exceptions that could form a reasonable basis for doubting that the playal brain is the basis of REM sleep. But, if this is so, I believe that there remains no reason for doubting that the whole development of copy-learning, and of the learned-routine space, and of understanding, and of "felt" selfness, and of REM sleep itself, grew out of the unique situation of the post-reptilian animal-eating animals that reared their edible young in close proximity to themselves. In the resolution of the associated paradox, these animals came on the remarkable and unique features of play. This provided precisely the conditions required for the emergence of the other developments just listed, and which could not have arisen without the unique contact and interaction that occurs between parent and young during play.

This is, as I am sure you realize, probably the most significant conclusion that will be reached in this book and so, even at the expense of some repetition, it is important to make the nature of the conclusion clear. We can begin by noticing that both mammals and birds display the capacity to develop modes of behavior that are not hereditary, that is, that can be shown not to have been transmitted by means of genes.

These modes of behavior reflect what has been called "copy-learning." As shown in the previous chapter, it is copy-learning that, even in its most primal form, was associated with the emergence of understanding and differentiated "felt" selfness, and that must therefore have been associated with the earliest evidence of what has become such dominant elements of behavior in us. Since the behavior associated with copy-learning is so different from that associated with hereditary transmission, I have assumed that it must reflect the emergence in the brain of a new space which has been called the "learned-routine space."

We are therefore faced with what seems to me to be one of the grandest questions that we can ask ourselves: What occurred in the group of animals that evolved from reptiles that could give birth to this totally new way of arriving at programs of behavior? Clearly, the answer is not to be found in the emergence of the mammal as a mammal, since copy-learning is also found in birds, and this is also expressed in the presence in them of REM sleep. So what do mammals and birds have in common, which could lead to the emergence of copy-learning and REM sleep in them both? We might presume that it is simply the fact of their common reptilian heritage, but the echidna, descended from reptiles, and lacking REM sleep, negates this presumption.

Indeed, it is not easy to find any common features between birds and mammals that would account for the sharing of behavior so particular as that associated with copy-learning and REM sleep. One suckles its young, and the other arranges for the feeding of young in a variety of ways all of which exclude suckling; one has fur, the other has feathers; one lays eggs, the other bears live young, although here, the echidna, like other monotremes, occupies a special pre-mammalian niche, for it lays eggs, bears "fur," and suckles its young; and so it goes.

So what do they have in common that could lead them to sharing the power of copy-learning? I believe the answer is clear, standing, as it does, in the light of what has gone before. What these disparate creatures share is the fact of having had ancestors that were faced with the ultimate paradox of being animal-eaters that reared their edible young in close proximity to themselves. The mere existence of such creatures testifies to the fact that they evolved a way of resolving this paradox, and the form that this resolution took is still evident in many of their descendants today in the form of the attenuated acts of killing that run beneath the ancient masks of "play" between parents and young.

Out of the uniquely close and extended contact between parent and young that is the essential nature of play, came the conditions for copy-learning, and what followed it. The pieces fit nicely, now—all the pieces, including the learned songs of birds, and the learned hunting behavior of some playals, and their sleep with REM passages, and tickling, and laughter. And we can see why those of them that have persisted and grown in the ways of animal-eaters while rearing their young had to become more determined and overt playing animals—the true playals of today; and that is why they copy and understand more, and have sharp REM periods in their sleep.

And we can begin to see why these playing animals have such a great and "unexplainable" attraction for us, an attraction that is so clear that we can only account for it with another piece that fits, and this is the piece that shows how very much we are like them, and they must be like us, in those many feelings that we can't but share with all play's children.

This is a good place to tidy up a loose end, left dangling since Chapter II, which concerns what can actually be seen as play in three plant-eating primates: the gorilla, orangutan, and gibbon. What is evident now is that, within the classes of birds and mammals, the boundary, between those creatures that will actually be seen to play and those that will not, is somewhat softer than the beginning picture of that chapter would suggest. Now we can see even more clearly why birds and mammals that are presently plant-eaters might show signs of play. This is due not just to the weak, vestigial playal quality that we can expect to find in some of the plant-eaters among post-reptiles, as mentioned in Chapter II, but much more to the effect that their capacity for copy-learning, as confirmed in them by the clear presence of REM sleep, might have on their external behavior generally, and particularly on the presence in them of copy-learned versions of play. This would seem to be particularly relevant to primates as close to us biologically as the three mentioned, and even more so when they are held by us in intimate captivity.

THE REALITY OF THE INNER STREAM

If you were to search the literature of neurobiology, you would not find a clear reference to what I have been calling the "inner stream." There is only one possible exception to this statement, and it is so famous that it would be neglectful beyond reason to leave it unmentioned, so I shall treat it at some length later. But in spite of its interest, even the exception that I have just mentioned is not really a reference to the kind of nervous stream that I have been using to explain the nature of copy-learning, and now of sleep.

What you would find in the relevant literature are references to "proprioceptors" and "proprioception," following the brilliant work (and the terminology) of Sir Charles Sherrington in showing how muscles report their states to the brain; but there is no reference to the role that the continuous stream of nervous signals flowing in this way must be playing in determining the endless dynamics of behavior, and, ultimately, in copy-learning. The lack of recognition of the need for such a role for this stream is all the more surprising since copy-learning is such a commonplace in our own lives, and is so clearly in need of explanation.

The need for an inner stream, which ultimately determines behavior, arises as soon as we realize that the only source of external behavior is some program of nervous signals that can actually be connected to the individual nerves and muscles of a real, individual creature. The primary source of such signals is the genetic endowment of each particular creature, and this provides the behavior typical of its species and, to a much lesser extent, the behavior typical of the creature itself. A genetic program can drive actual behavior because the same program that controls development also controls behavior, so that we don't have one creature's program for development trying to drive directly some other creature's behavior.

When we come to explain copy-learning, the same constraint applies, in that the only way in which we can have a program that can reproduce behavior is to find some way of storing the creature's own, actual neuromuscular behavior, so that the stored signals can literally connect back in to the same nervous system that generated both the behavior, and the signals, in the first place. When we come to see the situation in this way, it becomes clear that just images, flowing from outside the creature, of whatever kind, visual or aural or whatever, can simply never be a basis for repeating behavior, because there is no actual nervous stream associated with them that could be stored anywhere, which can be re-inserted into the nervous system of the creature, and bring on prescribed behavior. The most that images can do is bring on certain of the creature's own programs, which then determine its behavior; but the images themselves simply have no way of directly driving muscles.

This is the argument for the necessity, and the existence, and the reality of an inner stream, and its store. But, much more important than any argument of this kind is what happens if one tries to apply the notion of inner stream to situations in which the behavioral facts are known. What kind of model does this lead us to, and how does the model perform in comparison with the behavioral facts, as they are known?

Of course, it is precisely this kind of modeling and comparing that I have been conducting with the inner stream in this and the previous chapter, and, without wanting to be disrespectful of the invaluable work of others, it seems fair to say that the model of copy-learning, combined, as it has been, with that of sleep, both based directly on the inner stream, is the most complete and veridical that has yet appeared.

Further, by admitting only biological entities such as nervous streams, and stores that admit of identification and measurement, the model avoids lapsing into "explanations" that could have no evolutionary basis, and hence no permanent place in the explanation of the behavior of biological assemblages. I therefore assume that either inner streams and their stores, as I have been using them, are real, or that they are such a complete metaphor for what is real, that their use in the explanation of the behavior of living creatures is justified.

You have probably guessed that the exception mentioned at the beginning of this section referred to the work of Sigmund Freud. And well it might, since all of his pioneering work on sleep and dreams is based on a model with both an inner and an outer stream. The outer stream that I introduced in the previous chapter is the same as that of Freud's model, and is made up of the signals arriving from outside the creature which serve to define the environment in which it must continue to survive. However, my inner stream is different from that in Freud's model, and the differences can be seen clearly in two areas.

The first difference is probably just a reflection of Freud's interpretation of the neurobiology of 1895, which is when he laid the foundation of his work. This difference can be described by saying that whereas my inner stream carries *information*, in which the energy-levels involved need only be high enough for the signals to be distinguished from the surrounding noise, Freud's inner stream carries substantial quantities of *energy*, indeed, enough energy actually to issue directly in external behavioral activity.

The second difference is of another kind, which can be described by saying that whereas my inner stream is a flow of information describing the configuration of the creature, and is literally confined to potentially identifiable nerves, which have a well-documented biological existence, Freud's inner stream is a flow of energy powered by "drives," such as sex and hunger, and the paths in which these flows might take place are far from clear.

Given the differences just mentioned, and the fact that the ability of the Freudian model to describe behavior has, for some time, been held in question, it is useful to see if there might be anything in the differences that would lead one to expect that the present model might enjoy a better fate. I believe that the first difference, that between an information model and an energy model, is almost trivial, and could be resolved with relatively little re-working of the Freudian model. Thus, if the present model is to enjoy a better fate, it will not come from this difference.

But the second difference is much more fundamental, and here, I believe, are grounds for expecting a better fate. The reason for this is that the inner stream of the Freudian model simply has no basis in sufficiently elementary components of biology, and not because of any inadequacies in the neurobiology of Freud's time, but because the notions that underlie the inner-stream entities are too distant from measurable, verifiable aspects of biology, and of elementary behavior.

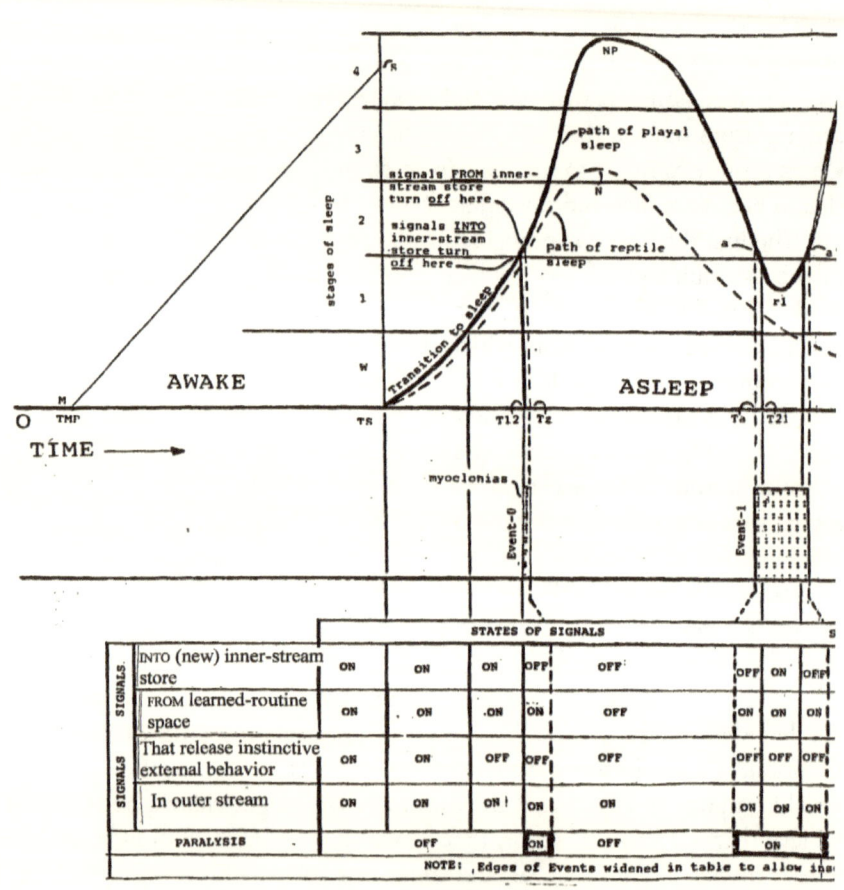

FIGURE IV. 3 a
PLAYAL SLEEP

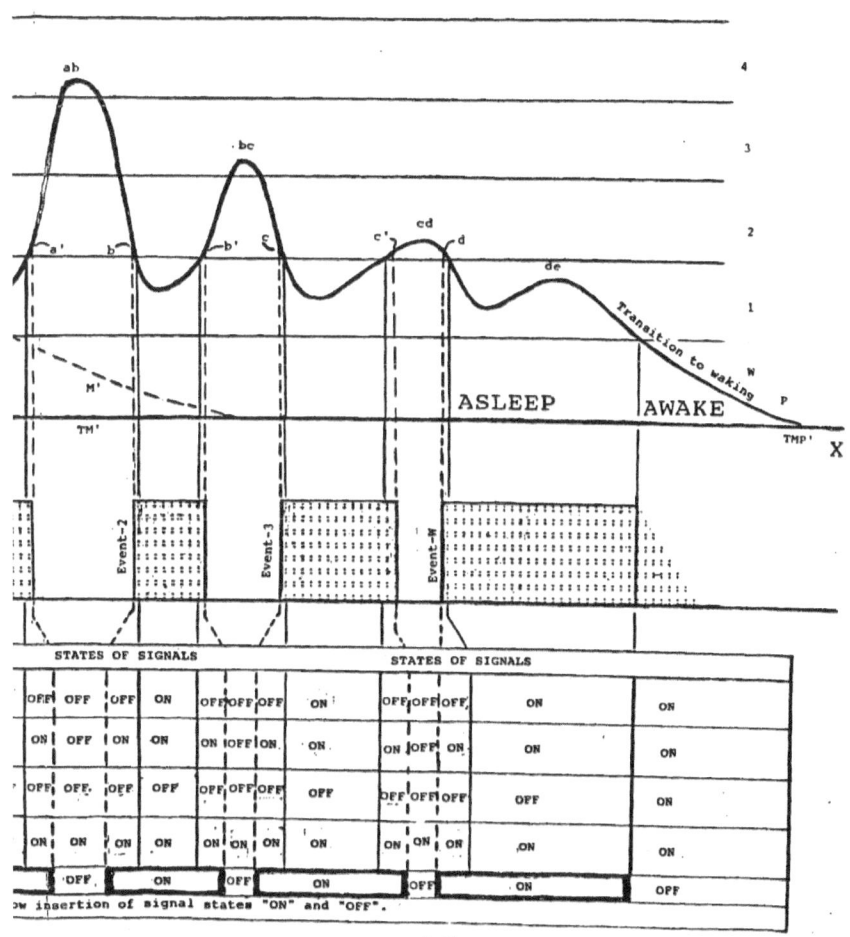

FIGURE IV. 3 b
PLAYAL SLEEP

It would seem that any model of behavior involving a brain must have at least two (actually three) streams, one inner and one outer, just as Freud would have us believe. But since the images in the outer stream will never be able to actually control the details of behavior, the inner stream must be so conceived as to be able to do this, or copy-learning will remain impossible. Without an inner stream that has this potential for control, the model will always lead to descriptions of behavior that are ambiguous at best, and unrealistic at worst, since it can never lead to any version of socially acquired behavior as the model's definitive prescription, and which can be compared with actual behavior, so as to be either confirmed or denied.

I therefore believe that the present model has the possibility of faring better than Freud's, because, as has been shown, in this and the preceding chapter, the model's inner stream leads not only to the explanation of behavior in a way that admits of comparison with actual behavior, but also to confirmation over a wide range of the creatures that we find around us today, when they are either awake or asleep.

But although the two models seem to take us along diverging paths, these cross again in a curious place which comes into view as we notice that the Freudian model has a certain haunting, almost sinister quality that links present behavior to intimate passages in the past. And we find that here again in this other model preaching that, if we are to find a route to a veridical explanation of ourselves, then it will have to come through viewing ourselves not just as mammals, nor primates, nor humans, nor as anything else just running in the evident present. Rather, such an explanation will lie in the expression of a long stream of evolution, which includes a passage that involved our ancestors in the resolution of a deep paradox that has marked what we are, and what we can ever become, with the indelible marking of "play." And the marking has spared nothing in us, not even our sleep.

EVIDENCE AND REFERENCES

Rather than increase the length of an already long chapter by adding numerous quotations, I have decided to mention briefly some of the books and articles on sleep that might be found interesting and relevant to what has been discussed. This also allows me to illustrate my deep indebtedness to a few of the many brilliant and dedicated people on whose work I have drawn, in order to acquire what little I have learned of sleep's excited tranquility.

For a simple but remarkably informative treatment of the whole subject, the little book by William Dement entitled *Some Must Watch While Some Must Sleep,* published by W. H. Freeman, remains unmatched. Comparable in simplicity and quality is *Research on Sleep and Dreams* by Gay Gaer Luce, written for the U.S. National Institute of Mental Health, and published by the U.S. Department of Health Education and Welfare. Much more recent, and beautifully illustrated, is J. Allan Hobson's book *Sleep*, published by Scientific American Library. There is a considerable amount of detailed information, with a great deal of emphasis on the "periodicity" of REM periods. There can also be found a number of theories of sleep that differ considerably from the one presented here.

Moving to more research-oriented material, the paper by Dement entitled "Sleep Deprivation and the Organization of the Behavioral States," in *Sleep and the Maturing Nervous System*, edited by Carmine Clemente and others, and published by Academic Press, is invaluable for its singularly clear statement of what really characterizes REM sleep. The paper by Sterman entitled "The Basic Rest-Activity Cycle and Sleep: Developmental Considerations in Man and Cats," even though it perpetuates the statement that sleep in infants "begins" with REM sleep, which I have critiqued at some length earlier, is otherwise full of useful information.

The book *Sleep, Dreams and Memory*, edited by William Fishbein, and published by Spectrum Publications, Inc., contains a great deal of interesting material, expressing many divergent views on the nature of sleep, and especially REM sleep. But a singularly valuable paper on the significance of the work of Sigmund Freud, on which I have drawn heavily, is that by Robert W. McCarley entitled "Mind-Body Isomorphism and the Study of Dreams."

SLEEP DISORDERS Basic and Clinical Research, edited by Michael Chase and Elliot Weitzman, and published by Spectrum Publication, Inc., contains much detail that is useful. The paper "Whither the Sleep Factors?" by Mendelson, Wyatt and Gillin summarizes the work and thinking on sleep-inducing substances. The paper by Prinz and Halter entitled "Sleep Disturbances in the Elderly: Neurohormonal Correlates", contains information on an aspect of sleep that has received relatively little attention. The paper on "Aging and Sexual Dysfunction in Man: Contributions from the Sleep Laboratory", by Karacan, Salis and Aslan, treats the subject of penile erections during sleep in great detail.

There seems to be no better collection of papers on narcolepsy and cataplexy than is available in *Narcolepsy*, edited by Christian Guilleminault and others, and published by Spectrum Publications, Inc. The article "Cataplexy" by Guilleminault is particularly interesting.

All of these together, for all their imposing height when stacked on a table, make up only a minute fraction of the literature on sleep and dreaming, and the sometimes vast bibliographies to be found in some of them will confirm this.

CHAPTER V
GAMES

The Most Primitive Game

If we think of the playal that has acquired the offensive components of primal play by copy-learning, then, as explained in Chapter III, more and more of its programs of predatory behavior will gradually be transferred between generations by means of such learning. In the early form of this development, the creature has two ways of executing the same behavioral routines, and the problem of action is solved by the routines derived from copy-learning having priority, because of their survival advantages. Eventually, however, some of the parallel routines based directly on genes and heredity disappear altogether. This occurs because of the tendency of natural selection to lead to the eradication of unnecessary functions in a brain which, especially in mammalian playals, is large and growing, and in which space is at an increasing premium due to problems at birth posed by a larger and larger head.

To see what this might lead to, it is useful to break the offensive aspect of primal play into three separate parts as follows:

A. The program that connects hunger to seeking and attempting to kill prey;

B. The program that drives the actual killing behavior that underlies primal play;

C. The program that limits the expression of the violence that underlies primal play to the harmless activity that it must remain.

We can assume that, if copy-learning were ever to assume the instinctive function in "A," it would do so at the very last, since A is so absolutely crucial to the playal's survival. So, if we assume that program B is running on copy-learning, what about program C? Although the consequences of the answer are enormous, as we shall see in this and the following chapter, the answer itself is simple, as are so many of the answers to evolutionary questions when we happen to stumble on them.

The answer is that C is not easily separable from B, and so C also shifts toward an implementation by copy-learning. As the shift develops, the suppression of violence that underlies primal play is achieved by the bodily expressions, mainly those on faces, which the playal carries over from the previously instinctive phase. In this way, the playal substitutes for a previously internal, instinctive, hereditary control on the underlying violent behavior of primal play, *two sets of external, expressive signals, one set acquired in defense and the other in offence, which serve to constrain the associated behavior, and limit it to play.* When this becomes anchored in copy-learning, the underlying instinctive controls progressively disappear, for the reasons outlined previously.

We arrive, in this evolutionary way, at a type of copy-learned behavior which resembles primal play, but in which the moderating factor that limits the activity to play is now rooted in facial and other expressions which must accompany the behavior, and serve to define its context. This kind of behavior is, evidently, what we refer to as a "game," by which we mean an activity in which at least some of the normal consequences of behavior have been suspended for as long as certain copy-learned expressions continue to constrain the range of behavior which forms part of the activity, that is, for as long as certain copy-learned expressions continue to define the "rules" of the game.

This kind of activity is clearly no longer usefully viewed as just primal play, but rather as an *artificial* form of primal play, which is, evidently, one order more artificial than primal play itself. I shall refer to this kind of activity as "game-play."

What we would witness in the case of two littermates involved in game-play is a pair of animals seeming to be involved in a fight, with each one going "inexplicably" on the defense from time to time. They might display facial expressions peculiar to this peculiar scene, the expressions on one serving, more or less reliably, to condition the violence of the other.

We should expect that game-play will lack the precision and definiteness of primal play, since there is a greater range of individual behavioral mastery, that is, of "competence" associated with copy-learning than with the relatively precise performance flowing from the behavioral prescriptions of genes. This potential lack of competence might extend to the learning of the moderating expressions and their significance, and would be evident as difficulty both in initiating and sustaining game-play, and the avoidance of real fighting.

It is important to bear in mind as well, that, at least at the stage being discussed, program "A" above is still at work, that is, the initiation of game-play is the expression of the instinctive link between being hungry, and killing other creatures. It would therefore not be surprising to find traces of the behavior that is associated with primal play mixed into performances that are primarily game-play.

OBSERVED GAME-PLAY

The game-play of animals has been widely observed and documented. I have chosen the quotations that follow to illustrate two points. The first is the fact that careful field observations confirm the behavior predicted by the evolutionary development just outlined. The second point is that no existing theory accounts for the superficial "social" character of game-play, thus leaving room for an evolutionary origin of the type just outlined. You will notice references to "social play," and this can be read, for now, as equivalent to "game-play." The quoted passages are long, but so relevant, that it seemed reasonable to make full use of them.

The first of the three quotations is as follows:

> "After repeated unsuccessful attempts at maintaining play, partners may switch to a different style of play in which postures or physical relationships used are less likely to elicit serious fights. They design a game. For instance, young female rhesus macaques, in whom play-wrestling breaks down more frequently than in young males, tend to play-chase rather than to playfight. (Symons 1978 a). One sand cat (Felis margarita) kitten, aged 12 weeks, at Brookfield Zoo frequently mounted its littermate and bit its neck soon after play bouts began; the littermate repeatedly vocalized, hissed, and tried to get away. After several asymmetric bouts of this type and after several unsuccessful solicitations by each kitten, the mountee jumped playfully into a hollow under a rock, of its own accord (it was not chased). The two kittens then playfully and silently angled at each other with their paws for several minutes. They became so involved in their game that *they reversed roles four times (one kitten charged out and the other fled in) while playing for thirteen continuous minutes.* As soon as one kitten reached the safety of the crevice, pursuit stopped and the angling duel resumed. No other play-bout during the one-hour observation period

lasted longer than a minute." (My emphasis) (Fagen, pages 399-400).

The second quotation is as follows:

"Although earlier observers, including Charles Darwin (1896) and Gregory Bateson (1955, 1956), identified extraordinary conventions governing social play, Stuart Altman (1962 b) first presented an actual outline of these rules. He noted that rhesus macaque play interactions appeared to terminate quickly "unless the 'games' that were played were 'fair' games," i.e. unless each individual had about the same chance of attaining its tactical goal (of "winning"). Fairness resulted if like-aged monkeys played together or if a monkey whose strength or dominance status exceeded that of its partner actually held back ("self-handicapping") by matching the intensity of its acts to that of its partner's acts or even by taking a defensive or subordinate role, fleeing from a smaller partner or falling down near it. *Moreover, monkeys at play used special signals to communicate readiness to play, to solicit play from a partner, and to maintain play.* (My emphasis)

"Subsequent studies of animal social play confirmed that these and other mechanisms served to reduce risks associated with play and to stabilize play interactions. By virtue of these social conventions animals at play avoid injuring each other and reduce the likelihood of misunderstandings that would cause play to escalate into potentially dangerous fighting. Of course, this is the case only when such stability is in both individuals' interests." (Fagen, page 395).

I find the third quotation particularly relevant to what I am here attempting:

". . . If play interactions and relationships appear more stable and more cooperative than would be expected from simple calculations assuming ruthless self-interest, it then becomes necessary to specify how such stability or cooperation could evolve by *natural (including kin) selection.*

"The idea that social play represents an extraordinary form of cooperative social behavior is an uncritically accepted tenet of the biological study of play. Extraordinary cooperative social conventions, a social 'playground fence,' seem to ensure play's stability and safety. A requirement that play shall not harm either the players themselves or their social relationship is said to be enforced both by participants' own restraint and by individuals' actions outside the play interaction, including parents, other juveniles and adults. That these stabilizing conventions exist is clear. *It remains to demonstrate how they serve the genetic interests of those individuals whose behavior maintains them.*" (My emphases) (Fagen, pages 394-5).

Here we can see quite clearly that Fagen is insisting on an "explanation" of social (game-) play, which respects a constraint that I mentioned in Chapter I. What he is asking is that the explanation of social play not include new notions drawn in purely to explain social play and nothing else; in particular, he is asking that the explanation show how game-play "could evolve by natural selection" and how "the conventions . . . serve the . . . interests of those individuals whose behavior maintains them." I believe that the evolutionary explanation that has been offered here meets these requirements; but, in addition, it suggests strongly that it is not possible to "explain" game-play without having explained both primal play and copy-learning.

Three Kinds of "Play"

In Chapter II we came on two types of behavior that often are called simply "play." The first was what I have been calling "primal play" (and sometimes just "play"); this is the interaction between an animal-eating, young-rearing parent and its young when the parent's killing program is attenuated. The second is what I have been calling "exercise-play," and is associated with the storage of excess food as increased physical fitness in any animal. We have now come on a third kind of behavior which is generally called "play," but which is an artificial version of primal play.

This yields a quite highly differentiated picture of (what is usually called just) "play" as made up of three distinct modes of behavior. To benefit from the view which we now have, I shall, on occasion, find it useful to distinguish clearly between these three modes. On such occasions I shall refer to "primal play" or "exercise-play" or "game-play"; but, when the context allows, I shall continue to use "play" for "primal play." The literature on play (used in the most general sense) makes frequent reference to "social play," and I take this to be almost equivalent to what I have called "game-play," although there are occasions on which it seems that there is overlapping with primal play, which also has the appearance of being "social."

Although it isn't essential, you might care to re-read the quotations in Chapter II. If you do, it will, I believe, be possible to notice the increased clarity and significance that the observations take on when viewed in the fuller context provided by both the three kinds of play just summarized, and the softening of the boundaries of those species which can be expected to display playal behavior, because of the vestigial effects of the original animal-eating behavior of all post-reptiles, as explained in the previous chapter.

MEANINGFUL GESTURES AND PLAY-BOUNDED GROUPS

We have now come on the emergence of "meaningful gestures" that can take the form of facial or other types of expressions, and which have the important function of attenuating violence in the course of game-play. I believe that these "meaningful gestures" have fundamental significance for language, and this will be looked at in more detail in Chapter VI.

We can also see in the use of gestures and the importance of their being "understood," so as to limit the damage in game-play, that a particular type of playal will tend to play only with those others which have evolved roughly the same "language" of gestures, beginning, evidently, with playals of the same species, and spreading out from there to only a limited extent. In this way, the other animals with which a playal will play games becomes limited to those that share its gestural "language."

As the playal moves further away from those groups with which it shares a gestural "language" fully, it comes finally on other creatures with which it shares no meaningful gestures at all. It is among these that the game-playing playal finds its most natural prey, since it is with them that no way exists for initiating game-play, and, hence, for attenuating the still genetically controlled behavioral routine that drives it to satisfy a need for food by seeking and killing other creatures.

GAME-PLAY AND WAR

This last genetically controlled routine is, of course, program "A," of the three identified at the beginning of the chapter, and questions naturally arise as to whether this program, as well, might migrate under the control of copy-learning, and what could be expected if it did. We need only look at ourselves to see that the program can, indeed, migrate under the control of copy learning, since, quite clearly, we are not driven, by an instinctive program, to seek prey and kill at each onset of hunger. Thus, in us, the instinctive program "A" has been replaced by a relatively complex, copy-learned, socially-transmitted program that links hunger to seeking food in a greater variety of ways than just hunting and killing prey by offensive action.

But there is a very, very far-reaching corollary to this ultimate release from the narrow, instinctive program for satisfying hunger, which, in our playal forebears, tied offensive killing to hunger, and nothing else. This corollary is lodged in the room which *our* copy-learned program provides for offensive killing that *has no remaining roots in any immediate need to satisfy hunger.* This corollary, of which we are the clearest expressions, is on display whenever our offensive killing is directed by those socially-transmitted waves of behavior that slam our learning and strength against the shifting targets on which we turn, from time to time, both within our species, and without. Of course, a degree of control on this, in us, as in other playals, should come from shared "language," in the moderating transformation to a game; but just within our own species we have, not a single "language," but a numberless multitude, and, without it, even more. And therein lies the origin of the singular human capacity to kill, for some socially-transmitted "reason," on the gigantic scale we know as **war**.

However, regardless of the doubts that one can raise regarding the evolutionary outcome of game-play as it relates to ourselves, there can be no doubt that it provided the opportunity for the emergence of a steady stream of beneficial modifications in the behavior of playals generally, as this relates to the vital needs of feeding and defense. We can see how this comes about by noticing that an endless series of errors will occur in the continuing stream of copy-learning that game-play engenders. It is from this series of errors that new modes of behavior emerge, providing natural selection with exactly the variety out of which it filters those modes that fit the survivors a little better for escaping the ever-present maw of extinction.

CHAPTER VI
LANGUAGE

GAME-PLAY AS ORIGIN OF LANGUAGE

Except for some number of us who suffer from specific afflictions, we can communicate by means of speech and hearing. The whole process is so natural and effortless that we tend to be unaware of the complexities involved, having left these to the specialists who study language and its related organic apparatus. And of these specialists there have been many, so there has been no shortage of study, measurement, analysis and theory-building. But in spite of the insights gained in all these ways, there does not exist any even moderately satisfactory theory of the origin of language which can account for its dominant features. It therefore seemed reasonable to see if the communication that is the foundation of game-play might form a basis for an evolutionary theory of the origin of language.

Without trying to define language precisely at this point, it seems reasonable to assert that it serves to allow one playal to affect the behavior of another that might be some distance away, without the use of direct physical contact. This amounts to asserting that, at least in the simpler uses of language, which we might expect to be associated with its evolutionary origins, what we refer to as the "meaning" of each of the exchanges in language, each "signal," is determined by the *behavior* that the signal brings on. This seems to me to be the only way in which we can attach any demonstrable significance to "meaning," especially in an emerging evolutionary context.

As soon as we associate language and meaning with determinants of behavior, it becomes possible to place on both of them an evolutionary constraint: language and meaning must operate in the self-interest of playals. Thus, any evolutionary theory of language must find some activity, to locate its origin in, which allows participating playals to benefit *individually* and *simultaneously* from the power of meaning and language. For a number of reasons that I shall outline here, game-play seems uniquely able to fill the need for an activity in which to locate the evolutionary origin of language.

GRAMMAR AS A GAME

We can begin to see this by noticing that, in addition to having substantial survival value and hence self-interest for playals, game-play derives this value directly from the behavior that certain (facial) expressions must bring on, that is, from what certain expressions must *mean*.

It is important to notice, in addition, that the two playals involved in game-play are involved in *complementary* rather than *identical* activity. If playals are to have a low-risk game, rather than a high-risk fight, then one of them has to be chased while the other chases; one of them has to be attacked while the other attacks; one of them has to be "bitten" while the other bites. This is important, as I shall show in a moment, because of its significance in accounting for the nature of grammar, and hence for an acceptable evolutionary theory of the origin of language which must, evidently, account for the nature of grammar.

This significance for grammar comes from the fact that, with rare exceptions, each *active* form ("voice") of a sentence is (can be) associated with a complementary *passive* form ("voice"), and from the relationship that the origin of this fundamental feature of grammar might have to the complementary form of game-play, with its roles of attacker and attacked; roles which constrain the (language) gestures involved to having a natural "active" and "passive" character.

But, although game-play involves playals at any one time in complementary roles, the nature of the activity is such that, during the time of the game *as a whole*, every playal must eventually occupy *both* roles, and so come to practice the meanings of both the passive and the active parts of the signals, that is, to display the appropriate associated behavior in their presence. Thus, although the roles of playals in game-play at any given time are *complementary*, over the period of a *complete game*, their roles are also *identical*.

This distinguishes game-play from mating activity, for instance, which, although it involves mating animals generally, and playals in particular, in a kind of complementary activity, does not allow for a complete reversal of roles. Thus, mating activity could never support the emergence of what ultimately becomes our kind of fully *reciprocal* language, because it does not, as does a game, demand full reciprocity of behavior, and hence does not provide a framework in which to develop full reciprocity of meaning.

Such bringing together of complementarity and reciprocity in the roles of animals in the *same* activity is unique to game-play, and is probably an expression of its deep rootedness in the artificial; it clearly adds one more reason for viewing game-play as particularly appropriate ground in which to expect that language might take root.

Finally, game-play has significance for grammar because the game involves, albeit in a primitive way, the framework of a "sentence" to the extent that it must include three "slots" which can be filled with something like "I," something like "YOU," and something like "ATTACK."

I have summarized, in Figure VI.1 on page 115, what has just been said, by sketching the signalling situation in game-play. The sketch serves to emphasize the fact that a single signal consists of two parts, one part coming from one playal and a complementary part coming from the other. We can identify the two parts as the *active* and *passive* parts of the one signal. The sketch also emphasizes the three-part nature of each of the two complementary parts of the one signal. We can associate the three parts—I, YOU, ATTACK—in a primitive way with the three parts—subject, object, predicate—of a fully evolved sentence.

It is of the utmost importance to notice that the "feeling" of selfness which is implied by the "I" in such a beginning context for language poses no problem for the playal since, as explained in Chapter III, the earliest playals would, as a concomitant of the development of stored inner streams, and their extension to copy-learning, have begun to develop the degree of "felt" selfness required. In effect, the "I" can be seen as associated with an identifiable aspect of an ongoing biological process in the playal.

A similar observation can be made regarding the "YOU," since, again as explained in Chapter III, a playal would have begun to "understand" its parent(s) and siblings as separate individuals, having begun to copy-learn their behavior as an early part of game-play. Thus, parent(s) and siblings would constitute differentiated entities in the environment of even an early playal, different from itself and separate from the general background of non-self. In addition, the primary activity of game-play consists of behavior that can be copied, and this, in the nature of play, would be "ATTACK."

But if the playal brain can support the requirement for selfness which the origin of sentence-based (grammar-based) language being advanced here demands, we can also conclude that, if the origin of grammar-based language is indeed as I am suggesting, then there could have been no such language before the appearance of the playal brain, with its capacity for "felt" selfness and its associated understanding. This serves to emphasize how tightly linked are various features of the playal such as predation, rearing of young, play, copy-learning, "felt" selfness, understanding, games, grammar-based language, REM sleep, and the socially-induced killing of members of related and unrelated species. This also throws some light on the necessity of finding a single framework in which they can be placed, all of them at once, if we are to begin to grasp the significance of any one of them. As you will doubtless have noticed, this poses some quite considerable problems, simply in explanation.

A way of partly resolving these problems in explanation involves the (frequent) use of isomorphisms, of which you will have noticed a number moving in the background as silent ghosts, and I shall just take a moment to re-emphasize why. In the present discussion, what I am trying to do is construct an evolutionary basis, which is really to say, a biological basis for language. However, as we speak about it, language appears to be one kind of entity, and biology appears to be another. So to get from one to the other, in a way that is explainable, I look for a number of characteristic aspects of language, which, in this case, are features of the sentence that can be matched with a similar number of aspects of the biology of the playal, and which seem to play analogous roles. This (in my opinion) cannot be accomplished in a single step, and so I have applied isomorphisms in three successive steps; in the first step, I go from "subject," "object," "predicate" aspects of the sentence, to "I," "YOU," "ATTACK" aspects of game-play; in the second step, I connect these to "selfness," "understanding," and "copied-behavior" aspects of copy-learning; for the third step I make use of the isomorphisms of Chapter III, which connect aspects of copy-learning to aspects of biology.

Since this is important, and tends to seem complicated, I shall compare what has been done to how you would get from a statement in English to the equivalent statement in Chinese, if you had only the following dictionaries: English-French, French-Russian, Russian-Chinese. What you would (have to) do is go from English to French in a first "isomorphism," then from French to Russian in a second, then from Russian to Chinese in a third. Of course, languages aren't quite isomorphic, but they are close enough to being so to illustrate what I have been doing, in order to get from "sentence" to "biology."

Returning now to language itself, I shall add more detail to the link between the active-passive nature of game-play and the active-passive nature of the sentence, and this will provide more depth to the picture of game-play as a site for the emergence of grammar and hence of sentence-based language.

ACTIVE-PASSIVE NATURE OF SENTENCES

Evidently, the signals in English are sentences, and the first thing we find is that, but for rare exceptions, all English sentences come in pairs, active and passive; and, when experts look at other languages, they find the same pairing in them all. In fact, this having an active and a passive form is such a general feature of sentences that Noam Chomsky, certainly the twentieth century's most renowned linguist, made it one of the cornerstones of his approach to linguistics. This is evident even in his earliest comprehensive treatment of linguistics in his book *Syntactic Structures*. Unfortunately, this book and the others by Chomsky with which I am familiar are extremely technical, and so, given the commitment to the use of everyday language in the one you are reading, they are not suitable for quoting here.

Fortunately, however, much has been written *about* the work of Noam Chomsky, and an especially illuminating short account is to be found in the book *Grammar* by Frank Palmer. In the following quotation the term "morpheme" can be taken to mean simply "piece of a word." (The book gives a full treatment of the history and significance of the term.) Palmer's discussion of the active-passive pair begins as follows:

> "In 1957 Noam Chomsky's *Syntactic Structures* was published. This was the book that first introduced to the world the most influential of all modern linguistic theories, transformational-generative grammar, TG for short. The theory was, undoubtedly, revolutionary, but as with all revolutions, some of it had already been foreshadowed in earlier works, particularly in the writings of Chomsky's own teacher, Zellig Harris, and even in the writings of some scholars who were and have remained within a theoretical framework that seems utterly opposed to that of TG. Since 1957 Chomsky has modified his views and has on the whole moved even further away from the American linguistics of the 1940's and 1950's, especially in the more philosophical aspects of his theory, but much of what is important can still be found in *Syntactic Structures* and an understanding of the linguistics of that time therefore requires reference to this book.
>
> "The name 'transformational-generative' suggests, quite rightly, that there are two aspects of the theory. The grammar it provides is both 'transformational' and 'generative'. These two aspects are not logically dependent upon each other, though the theory gains plausibility from this interaction of the two. But the two aspects can and should be considered separartely. The transformational aspect is the more

fundamental and perhaps more revolutionary and we shall, therefore, begin with that.

"Transformation

"The grammatical theories which we have so far been discussing were concerned very largely with the analyses of sentences in the sense that they must be divided into parts and that the functions of the various parts must be stated. This was not of course an end in itself. Analysis of this kind allowed the investigation to show how one sentence was related to another in that their descriptions would be partly alike and partly different. If, for instance, we consider the sentences *John likes Mary* and *John liked Mary* the difference consists solely in the occurrence of –s rather than –d, or, to put it more technically, of the present tense morpheme rather than the past tense one.

" . . . There are, however, plenty of sentences in English which seem to be very closely related, but whose relationships cannot be handled in this way. The most striking examples of this kind are the pairs of active and passive sentences such as

John saw Mary.

Mary was seen by John.

or

The teacher allowed all the little children to go out to play.

All the little children were allowed to go out to play (by the teacher).

"Not only are these pairs of sentences related, perhaps even more closely than the present tense/past tense pairs we considered earlier, but also we can talk with some plausibility of the passive sentences being 'formed from' the active ones or even of being 'the passive of' the active ones. Yet there is no way in which their analysis in terms of any kind of phrase structure grammar can show this relationship. Indeed there were some linguists who actually denied that active and passive sentences were related grammatically and insisted that the relationship was purely semantic, i.e. that they merely had (roughly) the same meaning.

"What we need then is a theory that will not merely allow us to replace one element by

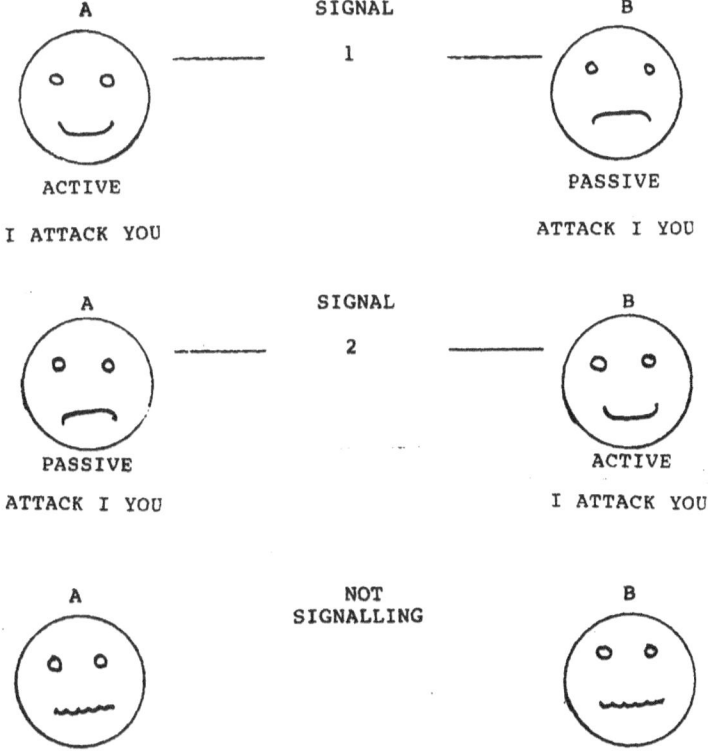

FIGURE VI.1

SIGNALLING IN GAME-PLAY

another or by a number of others, but also to take the sentence and completely rearrange it. We have to note then, in order to relate the first pair of sentences, that in the first of them (the active one) *John* comes before the verb and *Mary* after the verb, while in the second (the passive) *Mary* comes before the verb and *John* after the verb, preceded by *by*, and that this change in relative position is to be accounted for in the grammar. (In traditional terms *John* is the subject and *Mary* the object of the active sentence while *Mary* is the subject and *John* the 'agent' – perhaps, as there is no common technical term – of the passive sentence.)

"We can make a general statement about these relations. We can, that is, state how we convert an active sentence into a passive sentence: we have to change the position of the nouns or noun phrase and insert *by* before the second one in the passive and at the same time change the verb from active into passive. This Chomsky refers to as a 'transformation'." (Palmer, pages 135-7.)

I shall add to this long quotation one short additional comment by Palmer which occurs later:

"We do not want to justify the passive transformation by asking that the active and the passive have the same meaning. The essential point rather is the one that Chomsky made, that *an active sentence (with rare exceptions) can be transformed into a passive one*." (My emphasis) (Palmer, page 149.)

What is clear from even this brief summary of a part of the work of Chomsky is that any evolutionary theory of the origin of language that does not, as one of its integral parts, make room for the emergence of the two-part, active-passive nature of signals that go to form language will face insurmountable difficulty in being reconciled with one of the more fundamental features of existing languages and of modern linguistics.

So it seems to me far from trivial that game-play should provide at least the suggestion of an evolutionary explanation for something that people who study language have wondered about for a long time, namely: What earthly use the passive form of a sentence could have, other than to torment the people who study language. Indeed, what the situation in game-play suggests is the emergence of language in which the active-passive aspect of its grammar, as of Chomsky's transformational grammar, very nearly *is* the language, rather than just a striking feature of it.

As we come on the signals in game-play, they consist of two parts bound together in an active-passive pair. Each of these parts functions as a sentence and has a natural three-piece structure. Let us call these pieces I, Y, A to remind us that a playal signal has to have in it something I-like, something YOU-like and something ATTACK-like, or it simply would not have the basic capacity to convey what it must convey to initiate and sustain a game between two particular playals. It doesn't matter in what order a playal assembles these pieces, so long as it does it the same way every time. Let us assume that the order is I A Y for the active part.

Of course, we must have a passive part to go with the active part and so complete the signal. There are two basic ways in which we might construct it: one depends on changing the relative positions of the pieces I,A,Y; the other depends on making some minor little change in one or more of the pieces, just enough to show that we're looking at a passive form, but not so much that we lose track of which piece is which. But, since the business of being sure of who is doing what is really vital to the playals in game-play, the way that evolution could be expected to proceed would be to end up with a combination of the two basic ways, since that would provide, for some chance variety of playal, the most powerful way to reduce the numbers lost in fights that grew out of what, to work properly, has to remain play.

As soon as we introduce this range of possibilities, a whole host of ways turn up for constructing passives. However, I shall limit myself to just two ways of proceeding which I shall refer to as P (as in playal) and E (as in English). The sentence-pairs that make them up look like this:

P: I A Y   -   A I Y

E: I A Y   -   Y a I

Since it might not have been noticed, I call attention to the fact that the P-type passive is produced by a single operation in which we move one piece. In the case shown, the "I" jumps from the outside to the inside. In the E-type passive, we have to make two jumps, the "I" from one end to the other, followed by the "A," from the then outside to the inside. Another way of viewing this is that the P-type is a single *translation* of a piece, and the E-type is a single *reflection* of all the pieces in a mirror at the hyphen. So the two ways of forming these passives are really based on more than just change of position plus a little change in the pieces, but also include how many moves we make in arriving at the passive, or, what amounts to the same thing, what kinds of single operations we perform.

To illustrate why the E is as in English, look at the following sentence-pair which makes up a signal in English:

John loves Mary ----------- Mary is loved by John

I   A   Y   -----------   Y   a   I

That is the way in which signals in English are built. One might argue that we should take the *is* with *Mary* and the *by* with *John* and write the signal-pair as:

I   A   Y   ------------   y   a   i   ;

but this involves much more detail than I need to pursue, and, as important, will divert us away from Figure VI.2 on page 122, which is where we should now go.

This figure is intended to illustrate the fact that, using the features of a playal consisting of its mouth and chin, it is possible to form a P-type signal pair. What this suggests is that the playal is able to support a grammar consisting of an active-passive pair of single sentences consisting of three parts arranged in a form much like that of an ordinary English sentence, except for a comparative simplicity in the formation of the passive. (This is yet another isomorphism.)

I believe it is not only easy but reasonable to see in all of this the origin of language and the way in which it began to evolve. We can begin to see that there is an explainable evolutionary origin of what we refer to as "grammar," of which the parsing of sentences is the clearest expression, and, for some, the most unwelcome reminder of life at school. But the fact that we have sentences that can be consistently parsed at all, that they occur in active-passive pairs which have simple rules that bind the pairs, that they either have or can be understood to have three parts, that the sentence is always parsable if it "has meaning," all of these reflect nothing more than a driving evolutionary necessity. They are, in fact, nothing but some of the evolutionary consequences of richly self-serving game-play.

Evidently, the relatively complex passive to be found in English was evolved from a simpler form which can be executed in one operation with simple facial expressions alone. How we come to be using mirrors for our passives I leave till later.

There is an aspect of what I have been saying about the emergence of language that deserves particular emphasis here; this is the relation between the *structure* of the signals and their *meaning*, which is an integral part of the process of emergence from game-play that is being advanced. It is important to notice that a link between structure and meaning is essential in any evolutionary theory of language that is going to be able to connect to known languages, of which English is a good example, since it is clear that much of what a sentence-pair means, that is, much of the behavior that the sentence-pair brings on, depends on where in the sentence a certain piece of it is located, in relation to all the other pieces.

The importance and significance of this tends to be lost sight of in the present discussion because, when I use the face of a playal, as in Figure VI.2, to convey a signal, the relation between structure and meaning comes almost automatically from the fact that the parts of the face are placed in a certain relation more or less permanently, and that only limited rearrangement of them is possible. Thus, the fact that the nature of the process of language-emergence being proposed here is such that the link between meaning and structure comes about almost without insisting on it, tends to suppress an absolutely essential aspect of what is being proposed. The situation would be quite different if the emergence of language were being based on sounds, rather than on images, such as bodily expressions, and particularly those on faces.

SOUNDS IN LANGUAGE

For anyone unfamiliar with how the speculations on the origin of language have tended to run in the past, I should point out that I am, at this point, proceeding nearly opposite to the traditional direction. This is so because such speculation has almost always tended to start with the sounds, the "words" as we call them, which we utter when we produce the near-equivalents of what playals produce in image-sentences; near-equivalents that we can hear rather than see.

The most prominent and almost singular exception to this is found in the work of Gordon Hewes. The activity in which Hewes places the origin of language is tool-making but, in my opinion, the primary value of this approach was the attempt to direct the search at least partly away from approaches based exclusively on words.

It is easy to understand how the traditional search for the evolutionary origin of language could have come to depend so heavily on words since, for (most of) us, they are such an ever-present vehicle for language. But I hope it is now clear that, whereas pairs of simple images, arising quite naturally as an essential part of game-play, can constitute a basis for the emergence of language, there is just no imaginable version of pairs of simple sounds that can come even close to being able to serve as the origin of the active-passive form, for instance.

So if sounds are to get into the evolutionary stream of language, they will have to do so as late arrivals and be of some quite special kinds which happen to be:

1. Separable, so they can go into the three spaces that the structure of game-play provides in each half of a signal;

2. Able to go together so as to behave like the parts of an image;

3. Modifiable in ways that allow forming active-passive sentence pairs without becoming unrecognizable in the process.

It is the evolutionary origin of language, and the structural relations of the parts of the signals that accompany this origin, which determine the nature of the acceptable types of sounds--the "words"—not the other way round.

The primacy of pictures over sounds is such an important aspect of some of what I have just been saying, and much of what I shall say in what follows, that it would be reassuring to find evidence of this in humans today. As it happens, this kind of evidence can be found in the work of the late Eric Lenneberg. The following quotation sums up what a large part of his excellent book *Biological Foundations of Language* demonstrates in considerable detail:

[Note: What Lenneberg refers to as "language development" I have taken the liberty of paralleling with "development [of the use of sounds]."]

> "Relevant to this discussion are also two further types of observation. The first concerns the spontaneous play activities of the deaf preschool child. Unless there is also generalized neurological or psychiatric disturbance, the almost complete absence of language [the use of sounds] in these children is no hindrance to the most imaginative and intelligent play appropriate for the age. They love make-believe games; they can build fantastic structures with blocks or out of boxes, they may set up electric trains and develop the necessary logic for setting switches and anticipating the behavior of the moving train around curves and over bridges. They love

to look at pictures, and no degree of stylizing renders the pictorial representation incomprehensible for them, and their own drawings leave nothing to be desired when compared with those produced by their hearing contemporaries. Thus, cognitive development as revealed through play seems to be no different from that which occurs in the presence of language development [of the use of sounds]. On the other hand, and this is the second type of observation, mentally deficient children may have a much greater degree of language development [of the use of sounds] at age five than the peripherally deaf, but here the language [use-of-sounds] advantage does not help the level of sophistication for play." (Lenneberg, pages 362-3.)

But even if sound in language was an evolutionary late-comer, and today can be a victim without too serious loss in human function, it is clearly the case that sound has become the dominant vehicle for language in the vast majority of humanity.

That sound could become so dominant is not difficult to understand since, if we imagine a playal that is already benefiting from the use of image-language that works in daylight, its failure to work at night would be a handicap which some mutant gene would eventually reveal. Of course, all that natural selection needs is such a glimpse of advantage arising in some biological deviation, in order to unleash a series of extended probings with whole lines of animals as the probes.

But sound doesn't only solve the problem of darkness; it gets around corners, and through bush; it also works when playals aren't face to face, and when there are quite large distances between them. So it isn't difficult to imagine that, if a structural framework for language existed, such as that provided by the image-language of game-play, the power of sound, when placed within this framework, would ultimately lead to its dominance as the means of expression in language.

What might the emergence of sound-coupled-to-sight in language imply? Well, it implies a lot of things, and, to see what just one of these might be, we need to begin by noticing that the sounds in language must vanish quickly and follow each other in time, since, if they persisted, the latter parts of the strings of sound needed in language could never be realized as anything but long, uncontrollable passages of noise. Thus, the sounds in language that make up the parts of a sentence can never be continuously available all-at-once and together, as in the case of the parts of face-to-face images, spread out in space.

This is crucial, because the initial part of the signal, the passive part which goes out from the first playal inviting attack, needs to be stored and then checked as being a genuine call to play. Given the grammatical framework laid down by game-play, the way to do this is for the second playal to see if it can construct an active form of what it just heard and stored. If, as seems likely, both playals would have to do this all the time, each of them will need a brain in which it can carry the active and passive parts of every signal it hears.

This is not realized so simply as one might think, since the active part of a signal-pair induces one kind of behavior, and the passive part a different kind. Thus, if their external behavior is not to be fatally confused, playals can

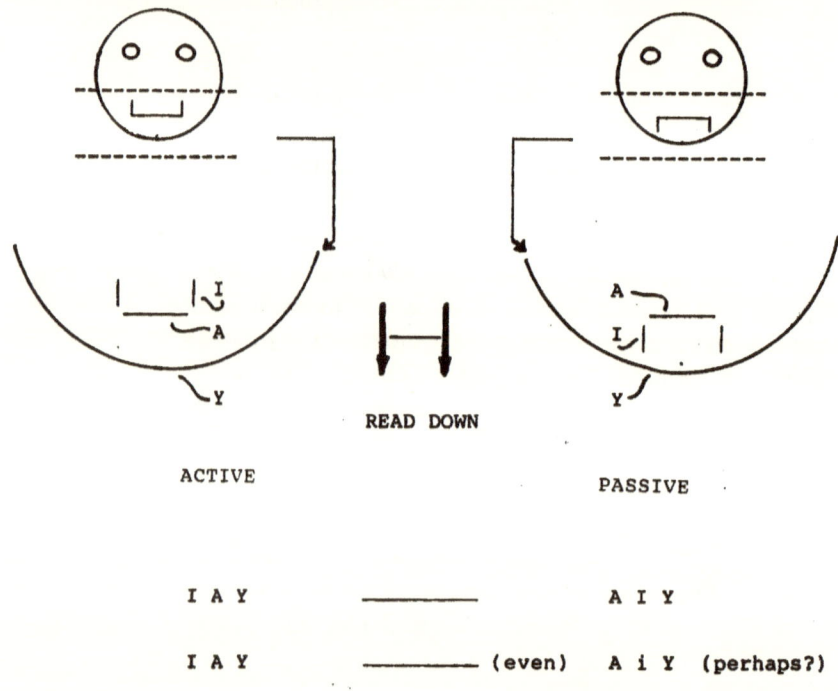

READ DOWN

ACTIVE

PASSIVE

I A Y	—————	A I Y
I A Y	————— (even)	A i Y (perhaps?)

FIGURE VI.2

A PLAYAL SIGNAL

advantageously evolve so as to include, in addition to any organs for generating the necessary sounds, a part of their brains where they keep active parts, and *another* part of their brains where they keep passive parts of language-sounds. In such a brain, the passive store would contain half-signals derived directly from sounds, while the active store would contain the complementary half-signals derived from the first store by means of the "logic" of the grammar of language derived from game-play.

Thus, to make use of language based on sound-borne signals, the playal will have evolved a sound-signal-related part of its brain with two separate but complementary "halves." One half stores word-strings in the passive form, and the other stores them in the active form. Since the passive is the invitational form for play, it will be stored by the invitee playal as a string representing directly what it heard; the active string, by contrast, will be a grammar-rule-based derivative of the other, and, to this extent, begins to take on a "logical," symbolic quality not present in the passive string.

Thus, to the extent that behavior is an expression of one part or the other of such a brain, it will display not only those features that are appropriate to one or the other role in play, but, in addition, will express differences between what is directly heard-and-stored in the one case, and what is derived logically from such a store in the other. These differences will remain a part of the biological and behavioral nature of a playal as long as the portions of its language conveyed by sound continue to rely on a grammar based on the active-passive structure of game-play.

As you will have realized from what was said a few paragraphs back, I am implying that humans are just such playals, to the extent that, in the use of English, for instance, they clearly make use of an active-passive grammar. I am therefore faced with the question of whether there is evidence of such separation of even a portion of our brains into two functional parts, one associated with the active "voice" of language, and the other with the passive. I think there is such evidence, and that it is part of the evidence for what is known, increasingly, as left-brain and right-brain function.

LANGUAGE AND LEFT- AND RIGHT-BRAIN FUNCTION

What I shall be doing shortly is citing evidence in support of the nature of the link that has just been proposed between two complementary parts of language, and complementary differences in function between two parts of the brain; but, before doing so, three introductory observations would be useful.

The first relates to the fact that a major portion of the (post-reptilian parts of the) human brain is arranged in two "similar" halves. These two halves are situated symmetrically on the left- and right-hand sides of an imaginary vertical plane passing through the centre of the head from front to back, and are generally referred to as the left and right "hemispheres." The two hemispheres are connected together by a massive neural bundle, called the "corpus callosum," which runs between them. The human is only one member of a large group of existing creatures that have brains with two hemispheres; and many of the other members are, in evolutionary terms, distinctly ancient compared to us. I now make the following assertion:

> The left and right hemispheres are the places in which the active and passive forms respectively are stored, and the necessary links between them are provided through the corpus callosum.

The second observation relates to the first, in that when I assert that the active form of language became associated with one hemisphere, and the passive form with the other, this is intended to imply simply that "separate" hemispheres constituted an ideal pair of places in which to begin to locate and separate the parts of signals, not that the hemispheres emerged as new parts under the pressure of language.

The third observation is of a quite different kind, and is directed at helping to interpret the evidence that will be presented shortly. An important part of this evidence is based on behavior observed when either one or other of the hemispheres, that is (what I consider to be) the location of either the active or the passive store, is severely damaged; and so it is useful to imagine what changes in behavior might be expected under such circumstances. What I wish to emphasize, in this introductory observation, is the fundamentally different kinds of behavioral deficiency that the foregoing assertion would lead one to expect depending on whether the active or the passive location is damaged.

So let us assume first that the passive store is damaged in the playal which is being invited to play. The starting situation then is that the playal wanting to initiate play sends out a passive sentence inviting play. When the invitee playal with the damaged passive store hears the passive sentence, it might be possible (because of the extensive interconnections and bypasses in the playal brain and in the human brain in particular) for it to lodge what it hears indirectly in its active store, produce an active form, and thereby appear to function normally since it will have, by being able to construct an active form of the passive sentence, "understood."

But the situation is completely different if we assume that the location of the active store is damaged, since, although the invitee can now lodge the passive sentence it hears directly in its intact passive store in the normal way, it can do absolutely nothing more with it since its *active* store is damaged, and it will now always appear to "not understand" *passive* sentences.

Thus, in summary, if the passive store (assumed to be in the right hemisphere) is damaged, the playal might appear to understand passive sentences; while, if its active store (assumed to be in the left hemisphere) is damaged, it will have no chance whatever of appearing to understand passive sentences. You will now see that such a striking and, in a sense, curiously inverted difference is worth noting and understanding if we are to be able to interpret the results of observations. I say "curiously inverted" since, if the passive store is damaged, an observer might conclude that it is functioning; while if the *active* store is damaged, an observer almost certainly would conclude that the *passive* store had a problem.

Evidently, if the degree to which hemispheres are judged to be associated with language is based on the apparent loss of function resulting from damage (and this is what is usually done), the active-processing (left) hemisphere will appear to be most strongly associated with language because of the absolute finality of the effects of damage in it.

Now I shall go to the evidence in the form of two quotations: the first is rather general and is intended only to reflect the question just mentioned of the degree of association of language with one or the other hemisphere; the second is specific and very directly relevant to active-passive storage and processing.

The first quotation is from the book *The Brain* by Richard Restak:

" . . . People may not only be of a 'different-mind' on issues, but they may also use different parts of their brains to do the same thing.

"This discovery tremendously enriches our understanding of the human brain. Instead of thinking that the left hemisphere is specialized for language, it may, more accurately, be specialized for symbolic representation. . . .

"Division of the hemispheres into symbolic-conceptual (left hemisphere) vs. nonsymbolic directly perceived (right hemisphere) avoids many oversimplifications. For instance, it is not totally true that the right hemisphere is completely devoid of language ability." (Restak, pages 250-2)

More direct and compelling evidence comes from the behavior of children who, as infants, underwent a "hemispherectomy," that is, a surgical operation for the removal of one of the hemispheres of the brain. It would probably surprise most people to learn that such children can display normal behavior over a wide range. The language-capabilities of three such children were tested by researchers Maureen Dennis and Harry Whitaker, as reported in the journal *Brain and Language*. A short summary of the tests is given in the book *Left Brain, Right Brain* by Sally P. Springer and George Deutsch, and I quote their summary (with some qualifying phrases of my own added in a few places, [and identified as such] to facilitate subsequent discussion):

> " . . . Maureen Dennis and Harry Whitaker have tested early hemispherectomy patients on various language tests and have found very subtle signs of lateralized effects [that is, effects related to different hemispheres]. Standard measures of verbal intelligence do not seem to differentiate between early left and right hemispherectomy. This does not mean, though, that other tests might not reveal such a difference.
>
> "Dennis and Whitaker studied three 9- to 10-year-olds who had undergone hemispherectomy by the age of 5 months. One was a right-hemispherectomy patient; the other two had had the left hemisphere removed. Results showed that both discrimination and articulation of the sounds of speech were normal in all three children. The three were also equally good at producing and discriminating words. Important differences between the hemispheres, though, appeared in tests of the patients' ability to deal with syntax – the rules for combining words into grammatically correct sentences. For example, each child was asked to judge the acceptability of the following sentences:
>
> 1. I paid the money by the man.
>
> 2. I was paid the money to the lady.
>
> 3. I was paid the money by the boy.
>
> "The right-hemispherectomy patient [passive damaged, active intact] correctly indicated that sentences 1 and 2 are grammatically incorrect and that sentence 3 is acceptable. The two left-hemispherectomy patients [active damaged, passive intact] did not make these distinctions.

"The researchers concluded that the right hemisphere [with its passive store intact] in the left hemispherectomy [active-damaged] cases *does not accurately comprehend the meaning of passive sentences.* Other tests led them to conclude that the right-hemisphere defect [in the left hemispherectomy, active-damaged cases] is *an organizational, analytical, and syntactical problem rather than one rooted in the conceptual or semantic aspects of language.*" [My emphases] (Springer and Deutsch, pages 194-5)

The results of these tests can be interpreted as confirming the earlier assertion that active and passive sentences are processed differently in different parts of the brain. More particularly, they can be taken as confirming that the processing to active sentences is carried out in the left hemisphere and the storage of passive sentences in the right because they confirm, in considerable detail, the expected consequences of damage to each hemisphere if one assumes that they are the sites of such storage and processing.

One should notice particularly that we can now shed some light on what is referred to in the last paragraph of the quotation as "the right-hemisphere defect" (in the active-damaged left-hemispherectomy cases), and this "defect" can now be strongly suspected as not being a right-hemisphere defect at all, but rather as the kind of absolute blockage of active-passive grammar and meaning that can be expected to result from destruction of the *active* processor and store in the *left* hemisphere, incurred unavoidably in *left* hemispherectomy.

So what is confirmed by these remarkably ingenious tests conducted by Springer and Deutsch is that to "accurately comprehend the meaning of passive sentences" the playal must execute the grammatical completion which can take place only if the *active* processor in its left hemisphere is intact!

I find it encouraging that conclusions which flow from the pursuit of the possible evolutionary extensions of play should receive such striking experimental confirmation in a species as far removed from the origins of play as we are. Of course, such confirmation is made possible because humans have become so remarkably proficient in the use of words. Let us, therefore, turn our attention briefly to them.

WORDS

As we have just seen, the evolutionary emergence of sound-based language leads to a number of identifiable biological developments in the playal. What I shall do now is pursue these a little further as they relate to words. A useful way to begin is by being reminded that, if sound-based signals are to find their way into language, they will have to be of a kind that can behave, to some extent at least, like portions of the images for which they must substitute.

The simplest portion of an image that we can find is the dot: . . (The first dot after the colon is the portion-of-an-image dot; the second is just the usual end-of-sentence dot.) The dot has, at the very least, three properties that we can see immediately:

> Diameter,
> Brightness,
> Position along a line.

The dot must have a large enough diameter to allow it to be visible to the naked eye, but, beyond this, its size can vary widely. Its brightness, by which I really mean its relative brightness, is what allows us to distinguish it from its background and surroundings. Its position along a line allows it to be placed somewhere in particular once the location of the line has been determined. It is important to notice that each of these three properties is independent of the others, that is, for any given dot, we can vary each property over a large range without having any effect on the others.

What we can see from this is that, if a simple burst of sound is to begin to make even the most limited penetration into the image-language business, and stand for even just a dot, it will have to have at least three properties which can, like those of a dot, be varied independently over some large range. As it happens, this is just what a simple sound-burst has, since it has

> Duration,
> Loudness (intensity),
> Pitch (tone, frequency).

These are, of course, not just three of a larger number of properties that a simple sound-burst can have, they are all the properties it can ever have. The trouble with this is that the dot is only the very simplest, ideal image we could wish to make a sound-burst stand for; there is no dot we can actually see which doesn't have some greyness or color. Consequently, in the world of even just one real grey dot on a white background, the simple sound-burst is already in trouble with respect to the extent to which it can stand for every dot.

The only way that anything more can become possible, in terms of its power to represent real images, is for the simple sound-burst to be joined up with others to make a single compound burst, for then, if the compound burst has even just two parts of fixed duration in which loudness and pitch can vary, the new part of the sound could represent the greyness of the dot.

As the compound sound-burst comes to have more and more parts added on, it improves its capacity to stand for more and more parts of increasingly complex images.

But there is something we can be sure of: there is no version of biologically feasible sound-bursts suitable for going into the spaces "I," and "A," and "Y," of a sentence, that will be anything but the vaguest representation of a playal's face, or a tree. So since we call one of these compound sound-bursts with all its internal juggling of frequency, intensity, and duration a "word," we can say with some assurance that there is no biologically feasible version of a word which will be anything but the vaguest representation of almost anything in the environment of the playal. So much so that one of them would later be heard to say (with words) in a burst of understatement: Even one of these old pictures is worth a thousand words!

You will have noticed that there is another ghost of an isomorphism roaming around, but now it fares rather badly, since what has happened this time is that I have used what seems to be the impossibility of setting up the exhaustive one-to-one match between aspects of sounds and aspects of images, which an isomorphism would need, to show that sounds can only approximate images, from which we can conclude that, in a very fundamental way, they are different kinds of entities altogether.

As indicated above, the way to partially make up for this difference is to use compound sounds of greater and greater complexity, but there must be a limit to the compounding, since useful signals can't take forever to be conveyed. Thus, the process is forced to a rough balance between vagueness of the words on the one hand, and the time taken for generation of the signal-sentence on the other. So vagueness remains, and vagueness always will remain, even in the most beautiful of those strings of words that we invent with such unthinking ease to describe the ever-changing world in which we are immersed.

As we all know, some of the most beautiful of these strings of words are embedded in what we think of as "music," and so a question arises as to the relationship between this "music" into which language can be so naturally embedded, and language itself. This is an interesting question, but it is most easily addressed in the context of a definition of language, which it should now be possible to attempt, given the earlier parts of this chapter as background. So I shall go first to the definition, and then to music.

A DEFINITION OF LANGUAGE

Even a not too careful observer would associate with humans a set of behaviors that could be grouped around what we refer to as "language." But it is easier to arrive at such an association than to formulate a definition of "the phenomenon" of language. However, using the foregoing discussion as background, one can construct a useful version of such a definition, and, in doing so, I shall begin with the definition of a fully reciprocal language, as follows:

(a) A fully reciprocal language is an evolutionary specialization that develops in a group of creatures.

(b) In a fully reciprocal language, the specialization in (a) takes the form of the ability of *every* member of the group to reproduce identifiable impressions in their environments, known as "*signals*," which either spread freely, or can be carried from place to place. Such signals include both transient images and sounds, as well as other less transient impressions in the environment.

(c) By means of the signals referred to in (b), one creature conveys to another a version of some aspect of what it has experienced. This kind of conveying of (necessarily past) experience between one creature and another is referred to as "communication," and what is conveyed is referred to as "information."

(d) The existence of a fully reciprocal language is demonstrated by showing that a particular aspect of behavior of every member of a certain group depends, in the same way as that of every other member of the group, on a particular signal when reproduced, as in (b).

(e) When a "particular aspect of behavior" such as that in (d) can be demonstrated, the behavior associated with the particular signal is said to be "the meaning" of that signal, and to "have a meaning" for every member of the group.

(f) The members of a group in which every member reproduces a number of particular signals, each of which, as in (e), has the same particular meaning for every other member of the group, are said to "use the same language."

(g) The members of a group, such as those in (f), for whom particular signals have the same meaning, that is, in whom particular signals were associated with the same particular behavior, as in (d), are said to "understand" the same language.

(h)   What might be called the "power" of a language is related to the range of aspects of behavior throughout which the existence of the language can be demonstrated.

It was convenient to begin with a definition of a fully reciprocal language since this (almost) corresponds to "human language," and so we can easily identify with it. But we should notice that such full reciprocity is not essential to language, and that there must be a vast number of non-reciprocal languages; indeed, although a human language is in principle fully reciprocal, in its actual workings, it never really is, since individual vocabularies differ. To adapt the foregoing definition to include such languages, as one certainly should, almost all we need do is, in all the statements (a) to (h) that are relevant, replace "fully reciprocal" with "non-reciprocal," and "every" with "some." In this way, we reflect in the definition the fact that, in a non-reciprocal language, not every member of the language group need be able to reproduce all the signals, nor does every member have to understand every signal. The peahen cannot reproduce all the signals that the peacock can, not having the same kind of tail, although she knows what the peacock means when it is grandly displayed; in this non-reciprocal way, the peahen understands the part of the language expressed in peafowl signals that only the peacock can reproduce.

We can conclude from all of this that the power of the fully reciprocal, sound-based, active-passive grammatical language of human beings is so enormous, compared to that of any other language, that it almost justifies the common title of just "language," implying, as this does, "the only language." However, it is probably one of the strengths of the foregoing definition that it allows us to avoid this extreme position, which would clearly be based on an excessive preoccupation with human function. Instead of trying to show that a certain way of communicating information is or isn't "language," we simply admit that there are languages other than "the (our) language," but that they differ in power.

As you might be aware, much ink has flowed in the cause of "proving" that human language is the "only" language, and somewhat less in "proving" that it isn't. What is clear is that our language must have evolved out of earlier languages, and it would be likely indeed that, in the course of this evolution, many streams would have emerged of which only some exist today, and, among which, human language is simply one. Equally clear is the fact that there are, all around us, purely instinctive languages, which preceded our partly copy-learned variety. These sustain communication between at least some of even the simplest creatures of which we are aware, among which insects are doubtless the most dazzling in this respect.

A much more interesting question than what is and isn't language is that of how human language came to develop the apparently limitless power that it displays. There are many parts to the answer, of which the first is undoubtedly the link between *structure* and *meaning* as expressed in the sentence and its grammar. Without this link, language remains a group of un-relatable alerting expressions having relatively limited power to affect behavior; but, with it, language begins to acquire the power to generate meaningful groups of images, followed by groups of sounds, since it becomes possible to insert sounds into the standard grammatical structure, and so produce an almost unlimited range of meaning based on the spoken, single-sentence active-passive pair. There is no such thing as language with the power of human language that lacks the structure of grammar.

In addition to this structural necessity, which seems to be at least partly embedded in the biology of the playal as "universal" active-passive grammar, I believe that the answers lie in the utterly unique circumstances of game-play. Everything about game-play suggests active-passive language, but probably nothing is so compelling as the fact that it provides for both a *complementary* and a *reciprocal* role for the playals. This goes to the heart of the roles in human language which must, over short periods of time, be complementary, and, over long periods, reciprocal; (most of) human language is, in its workings, both unsymmetrical and symmetrical, and only game-play, in all its full artificiality, could have been cradle for an infant language with needs that are so unreal and so essential, all at once.

Before discussing the relation of music to language, as promised earlier, I shall return to paragraph (g) of the definition, and begin by calling attention to what is referred to there as "understanding" a language. What I wish to show is the relation of this version of "understanding" to that advanced previously as based on copy-learning. The point to focus on here is the fact that "understanding," as we come on it here in a definition of language, is based on creatures having performed copies of some reference behavior, in the presence of particular signals. What this says is that the "understanding" based on the copy-learning of behavior, as we came on it in Chapter III, is really just a late-arising evolutionary extension of the more primitive "understanding" that must form part of all instinctive languages and the communication that they support. Looked at in this way we can see that the instinctive languages defined by genes and which, evidently, include "understanding," rely on copies of behavior transmitted by genes. This kind of "understanding" is clearly the precursor of our more extensive "understanding" which includes copy-learning in a way that extends understanding beyond behavior related purely to early communication and instinct.

In summary, understanding, of whatever kind, always entails some form of *copying* into a creature of some form of reference behavior. In the case of the playal, this copying is extended by the copying, into a brain, of behavior arising *outside* it, unlike the collective case of its predecessors, who were limited to repeated copying of behavior arising *inside* them, as prescribed by their genes.

MUSIC AND LANGUAGE

Given the rather broad definition of language just formulated, it would be surprising if music fell entirely outside it. What would seem to be the case, however, is that, compared to our active-passive grammatical language, music is a highly specialized sound-based language capable of communicating and affecting only what we experience as "mood." But even though the meaning of "mood" might be vague, we can identify it in reptiles, and amphibians, and fish, and so we might conclude that if music is a language that communicates "mood," it might itself be as primitive as these forms of life. I believe that it is, and that the croakings of frogs and lizards are just two of the myriad examples that one could give of the primitiveness of music.

Indeed, so primitive is music that it must have preceded primal play. As game-play emerged, active-passive grammatical language appeared in a primitive image-form which already contained all the rudiments of meaning related to structure. This provided the framework in which sound-based active-passive grammatical language could develop. The sounds of such evolving language must, at the start, have been the sounds of music, and its power grew out of the link between meaning and grammar. Because of the natural partitioning of the signals of the language of game-play into three slots, it became possible to put sound-bursts into the slots and produce a word-based language having meaning based on grammar.

To get a little better picture of all of this we can notice that when we hum a tune we are generating sound-bursts which are characterized by

> Duration,
> Loudness,
> Pitch,   and
> One Other Property.

The other property is a regular timing scheme according to which the sound-bursts must be emitted. This timing scheme we refer to as "rhythm." To see this clearly, hum the tune of *Twinkle, Twinkle, Little Star* to its rhythm which goes:

```
da da da da da da daaaa da da da da da da daaaa
tc tc tc  tc tc  tc  tc tc  tc  tc  tc  tc  tc  tc  tc  tc
1  2  3   4  1   2   3  4  1   2   3   4   1   2   3   4
```

I have added the regular tic, tic of a metronome to emphasize the regularity, and some numbers to avoid confusion.

Compare this with reading the following sentence in a natural way:

I am reading this long sentence in a very nat'ral way.

Although something like rhythm is preserved, the regularity is largely lost. To see this clearly, sing it to the constant rhythm of *Twinkle, Twinkle* and you'll see the difference.

This illustrates that while rhythm is a powerful means of expressing mood, and so is a vital part of music, it is far less important for our kind of spoken language. It would therefore appear that, as spoken language emerged out of the necessities of play, it took from music what it needed, and gradually shed those properties which were less important. Indeed, carrying rhythm as a basic property of spoken language would limit it to a kind of rigidity that you can notice by comparing the feel of the first three words in the sentence when read naturally together, with what you get when you separate them as you have to do to sing them to the rhythm of *Twinkle, Twinkle*, for example.

Spoken grammatical language thus came to depend heavily on pitch, duration, and loudness of the sounds making it up, and less on rhythm, which remained an ideal, but narrowly specialized medium for expressing mood, probably being connected to it in fundamental biological ways via a link to the rhythmic functioning of heart and lungs. In this way, the stream represented by the sound of early music split into two streams, of which one flowed into a series of elaborations as music, with rhythm as a fundamental and essential component. The other flowed into the processes of game-play, ultimately issuing in word-based grammatical language which still displays a strong rhythmic component in the form of poetry, but which is only weakly rhythmic in its prose form.

It is interesting to pursue the stream of music as music just a little further, since we then come on two extended strings of words which resemble passive forms of sentences and their actives, in the form of melodies and their harmonies. Further, the resemblance of the two situations:

(a) Someone hums a melody with someone
else who hums a harmonizing, counter
melody,

and

(b) A playal plays its defense in a game
as another plays its attack with subdued
violence,

is inescapable. But this, in my view, reveals nothing more than the frequent evolutionary tendency to converge on superficially identical "intersections" (as viewed by us), though proceeding along quite different paths, at different levels. In the case of music as music, evolution has led to a capacity to support an elaborate but rather restricted (orchestral) "game" based on the original benefits (necessity) of sharing moods, that is, of sharing the behavioral effects (meanings) of rhythms riding on sounds; in the case of spoken language, evolution has led to an elaborate process capable of supporting the most general of games based on the benefits of sharing the meaning of images as this must be done in the situation of game-play.

Thus, music has its meanings, but never speaks; since its language cannot speak. But speaking and music can be put back together as when we sing in words and music. The songs of humans are, in my view, the most powerful signals that playals can convey, and the reason almost certainly is that, while spoken language can speak about mood it can't do so while it is speaking about something else. But, when language is combined with music in song, we can speak of some subject and express an accompanying mood all at once, thus producing a combined effect which, fortunately, almost all of us can experience. This remarkable power derives precisely from the fact that music and word-based language are different rather than the same.

Evidence? Listen to Gregorian Chants, or the Beatles' *Yesterday*, or Spirituals, or *The Last four Songs* of Richard Strauss. Just listen . . . and cry, perhaps.

Or, imagine a movie without the music that accompanies it.

Of course, music is not the only language with which we can combine spoken language, since, to a greater or lesser extent, we all combine words with the waving of arms, winks, nods and whatever else serves to help us communicate.

But in spite of all this talk of language, you might have noticed that there is a very striking feature of spoken language that has not even been mentioned. This has to do with the evident fact that, in actual discourse, or in text as you are reading it, there is little evidence of the *passive* form of the sentence. How can it be that what must have been such a crucial element in the expansion of language in playals has come to play such a noticeably minor part in its present human processes? The answer to this question comes in Chapter VIII, where it emerges from an exploration of the evolutionary origin of *storytelling*. You might wonder what the evolutionary origin of such a commonplace could possibly yield of interest, but I leave the answer to that until we get there. Let us, before doing that, look at another aspect of play, and what flows out of it. Let us see what happens when a playal begins to challenge itself.

CHAPTER VII
THE ALMOST-HUMAN PLAYAL

What This Chapter Is About

The playal brain, according to the evolutionary picture developed so far, consists of two basic parts: one, which it shares with the brain of reptiles, relates to instinctive behavior, and another, which distinguishes it from the brain of reptiles, supports copy-learned behavior, and is the part in which are stored the inner and outer streams. These streams are derived from the episodes in primal play and subsequently game-play, both of which involve situations where the playal is challenged externally. Thus, the behavioral routines that can be generated from such streams are limited to those arising in episodes that involved *external challenge*, and so such routines, nothwithstanding their copy-learned origin, tend to resemble instinctive routines, in that they are linked to immediate, survival-based reactions to challenge.

But we know that, although many of our own behavioral routines are based on episodes which contain external challenge, we also display other routines which do *not* derive from such direct challenge and threat. So this chapter is about the evolutionary development that releases, from the limitation to challenge-based behavior alone, the part of the playal brain that makes copy-learning possible.

What we will find is that not only is there a quite justifiable evolutionary extension of what has been presented in previous chapters which leads to the required enlargement in the range of behavior, but also that this enlargement leads to relationships between the processes in the playal brain that allow us to distinguish its "unconscious" from its "conscious" life. This allows the evolutionary processes being developed here to be brought into correspondence with Freudian insights relating to unconscious processes in the human brain, but it is also shown that this kind of unconscious activity is just one segment of a whole spectrum of unconscious activities, flowing from a very long series of evolutionary developments.

In order to move the discussion further toward almost-human creatures, it is narrowed down from the generality of the playal (including birds and marsupials) to the consequences for just the (placental) *mammal* of the need for much increased memory and hence brain-space that must accompany the enlargement in range of behavior introduced previously. This is shown to lead to further changes in mammal behavior, including the disappearance of some instinctive functions and the need for a lengthening period of dependence in the young.

The discussion leads finally to very nearly the evolutionary completion of the major functional partitioning of the human brain, as it is now generally considered to be. Evidence of such partitioning is given, as it flows out of relatively recent "neurobehavioral" research, linking parts of the brain to certain kinds of behavior. Much of this research and its interpretation have originated with Paul MacLean and his colleagues.

THE PLAYAL CHALLENGES ITSELF

We can begin by going back to an observation made in Chapter III. This was that, during the episodes from which copy-learning and the learned-routine space began to emerge, the biological conditions were those that accompanied the intense challenge inherent in primal play. I then assumed that any subsequent access to the learned-routine space of a playal would require the same biological conditions.

This requirement to re-establish the biological conditions that accompany challenge, in order that signals might access the learned-routine space, has been met along the major lines of playal development discussed so far. It has been met in the case of game-play by its direct relationship to primal play, and the challenge that continues in the "game"; and it was met in the case of active-passive language by its emergence directly out of game-play with its associated challenge.

But we should notice that, in these cases, the requirement has been met by means of *external* challenge, arising and clearly visible in the external situation of the playal, whereas the new path of development, which I am now wishing to explore, stems from the possibility that the playal has for setting up the required conditions of challenge by an *internal* rather than an external process. The basis of this internal process resides in its stored outer streams.

Evidently, the earliest versions of the outer streams that are stored in the learned-routine space, with the inner streams that go to make up copy-learned behavior, begin by depicting the external *challenge* episodes of primal play. Thus, if the playal could "re-run" these internally stored outer-stream depictions of external challenge, and affect its biological condition in the same way as did the *original* external challenge, it would have a way of opening up its learned-routine space to new external experience which would not itself be constrained by the necessity of involving external challenge.

Now, although the previous chapters have placed almost exclusive emphasis on the capacity of the inner-stream store to replay stored sequences and reproduce external *behavior*, there is no reason to believe that the outer-stream store would be any less able to reproduce its contents and produce "images," including "visual" images. Certainly, the experience of dreaming supports this.

Further, one should recall that the entire process of copy-learning, as explained and applied in previous chapters, stems from the observation that, while image-sequences gathered in outer streams can *condition* behavior, by affecting various aspects of the general biological constitution of the playal, they *cannot actually drive behavior*. It therefore seems reasonable to suppose that, even in the *waking* state, the images residing in the outer-stream store could run *independently*, and come to situate the playal in a state of *internally* generated "challenge."

As always, any evolutionary development of this kind must support its biological costs with benefits, so it is convenient to be able to identify the benefits quite easily. As indicated above, the ability to set up a state of challenge internally, while awake, allows the playal to free itself from the copying of only those behaviors that relate directly to external challenge and immediate issues of survival, and so allows it to begin to include in its repertoire of copied behaviors some which, only later, might prove to be applicable to survival.

This kind of "experimental" behavior is what we associate with "curiosity," that is, a sort of general openness to new modes of behavior which have no immediate relation to survival. However, in the evolutionary scheme of things, curiosity has its risks, and it is important to see how these are moderated. This evidently takes place via a process of pain-labelling of past behavior, based on which the playal comes to re-use only those behavioral routines that were associated with minimum damage to itself, that is, with minimum pain, as compared to the other routines available to it.

Given this kind of moderating control, and even if, in the process of arriving at behavioral routines worth re-using, curiosity does kill some cats, the benefits to the playal of the expansion of its behavioral routines that come from setting up a state of challenge by an *internal* process, based on the re-running of outer-stream images, while awake, are clearly substantial. Consequently, we can expect to find the process of natural selection filtering the mutations, which continually occur in successive generations, in such a way as to lead to the gradual development of this process of *internally* generated challenge, based on the re-running of outer streams.

In summary, then, we have here come on an evolutionary basis for the appearance, in the brain of the playal, of what we might call "background challenge programs" that run while it is awake, and which serve to render it receptive to placing segments of its general external experience in its learned-routine space, in which it had previously been lodging only its primitive, directly play-derived external challenge sequences.

Indeed, we have here the evolutionary basis of a process that involves image-sequences running inside the learned-routine space of the playal's brain, which it need not "feel," or need only "almost feel," since the sole function served by the running of these sequences is the inducing of internal biological conditions similar to those arising in the presence of actual external challenge. That is, the running of these internal programs need not be "felt," as with copy-learned programs that support the execution of external behavior accompanied by "felt" selfness, since, to bring on biological conditions similar to those associated with actual external challenge, the internal-challenge sequences need only operate on the same *instinctive pre*-playal segments of the brain as those involved in the *instinctive* biological reactions to actual attacks such as those found in creatures *prior* to the evolutionary emergence of attacks in primal play.

Viewed a little differently, we can begin to see, in the development of internal challenge in the playal brain, an evolutionary basis for the emergence of a phenomenon that we should call "the unconscious life" of what is still a pre-human brain. In this way, we extend backward, in evolutionary order, the insight of Freud, who was the first to point to the existence of an "unconscious life" in the purely human context, when laying the foundations of his psychoanalysis. We should also notice that, in order to realize the necessary biological conditions of challenge, such an "unconscious life" would link a process involving image-sequences stored in the late-developing learned-routine space, of the playal brain, to its more *primitive instinctive processes*. Hence, such an "unconscious life" might be expected to display links to instinctive behavior, driven by the need to secure food and realize effective defense, as well as the pressure to mate, with all its sexual connotations—confirming, further, Freud's insight.

It is important to notice, in all of this, that there could not have been an opening up of the behavioral range of copy-learning, that is, of understanding, to the full experience of the playal without the emergence in it of internally generated challenge, and the associated unconscious activity in its brain. Further, this link between the necessity of unconscious activity and general understanding flows directly from the biological conditions attending primal play, mentioned earlier, in Chapters II and III.

There are a number of issues that grow out of this new process of internal challenge, which, as usual, provides new paths to further development as well as to extinction. Of these I shall now look at the following in some detail:

1. The existence of a spectrum of levels of "unconscious life";

2. The episodic nature of the contents of the learned-routine space;

3. The need to limit what is stored in the brain;

4. The effects of the increased need for brain-space on the development of the playal generally, but the mammal in particular;

5. The connection between internal challenge and inappropriate behavior.

THE SPECTRUM OF LEVELS OF UNCONSCIOUS LIFE

Although the unconscious life that I have been discussing must have links to instinctive levels in the brain, it emerges from processes that are tied to copy-learning and those levels of the brain that support it. The latter are levels that are particularly active in us, and provide the basis for felt selfness and understanding. Thus, the unconscious life that has such an emergence has a decided proximity to what we call our "conscious" life, and might be imagined to be "just next to it."

It is therefore interesting to notice that there must be a more remote level of unconscious life that is confined entirely to instinctive processes, which is clearly present in us, and which must have preceded the kind of "almost felt" unconscious that has just been discussed. All we need do to see this is consider that we can live an entire life without knowing how such organs as livers and kidneys work, or even that we have them at all. Thus, there is clearly in us, as there must be in the reptiles that preceded us, and in the creatures that preceded them, an extensive *unconscious* life, which proceeds at the level of purely hereditary functioning of much of our biological processes. Indeed, looked at in this way, we can see that the sort of "almost felt" unconscious that arises from copy-learning, and associated post-reptile playal segments of the brain, should be regarded as simply a late-arising form of unconscious, that is, of un-felt biological activity, which accompanies extensions of copy-learned behavior.

Proceeding in this way, we can imagine another variety of unconscious situated between the kind of primal unconscious just mentioned (which relates to development, in the sense of growth and function of parts of a creature), and the "almost felt" kind associated with copy-learning. Such an unconscious would be associated with instinctive behavior, as distinct from simply hereditary development, and again we can find such an unconscious in ourselves, and in creatures that preceded us. We can see it manifested in every unconscious blink of an eye.

In this way we can come to see that there is a whole spectrum of unconscious life, stretching from the most primal of biological activity, to that growing out of the part of the brain associated with copy-learning, and which allows the playal brain to finally place all its experience within the framework of copy-learning.

Viewed against this kind of evolutionary background, it is not the unconscious life that is the evolutionary stranger, requiring explanation, but rather the processes that provide the overlay of felt selfnesses, which we call consciousness. For, as we can see, it is not the unconscious that is the evolutionary exception, and that merits the prefix *un*, but consciousness itself, especially in us, for which we should have a word implying late arrival and exception.

EPISODIC CONTENTS OF LEARNED-ROUTINE SPACE

In Chapter III, I observed that the contents of the learned-routine space were made up of play episodes and their derivatives, and that this episodic character of its contents marked an important feature of the learned-routine space, and limited what it would ultimately be able to accomplish for the playal. Indeed, as I shall show in a later chapter, this determines and limits much of the human activity we call "science." So it is important to notice that the release of the learned-routine space from the storage of behavior related exclusively to external challenge does not alter this episodic character of the contents of the learned-routine space.

To see why this should be, we need to notice that, in order to produce the kind of behavioral extension associated with self-challenge, the outer-stream episodes that the playal brain had begun copying and storing during primal play must "switch" the receptivity of its learned-routine space "on" and "off." Thus, from the very start, this "gating" on and off of access to the learned-routine store is accomplished with earlier episodic scenes having a definite duration. Thus, the episodic structure, with which the learned-routine space begins, continues to determine its own ongoing structure, and, although the lengths of episodes might increase slowly until they become much longer than the initial episodes of primal play, the gating of the learned-routine space by its own contents, which frees it from the limitation of its beginnings in purely external challenge, will continue to maintain the episodic structure that existed at its early beginnings.

LIMITING WHAT IS STORED IN THE BRAIN

By comparison with reptiles, playals have a fundamental problem relating to the storage of those contents of their brains that are associated with ongoing life experiences. This is so because, whereas a reptile can be regarded as fully programmed genetically, and therefore not in need of additional memory in which to store its life experiences and what derives from them, a playal is distinguished, in large part, by its capacity to copy whole programs of *new* behavior that it encounters in its environment, and this must require a relatively large brain-space in which to carry out the necessary storage.

Added to this, and as just explained, the playal is eventually freed of its dependence on external challenge to allow access to its learned-routine space, and this aspect of its development adds enormously to the need for brain-space in which to store the substance of this complete exposure of its brain to the playal's full, ongoing existence. Viewed in this way, the problem of space in the brain, in which to store images and derivatives of current experience, comes to be not just one of the problems posed by the emergence of the playal, but very nearly *the* problem.

The problem of storage consists of two related parts; one of these has to do with the resolution, and the other with the extent of what is to be stored. The problem of "resolution" is easily appreciated if one thinks of a face, for example, that is to be represented in some kind of memory. It can be represented as an oval blob, at one extreme, or as a carrier of great quantities of detail, with every wrinkle and pore preserved. I refer to this as a problem of "resolution," since it has to do with the level at which the brain is to "resolve" the details of the face. Evidently, the greater the resolution, the more the amount of memory a brain is going to need to represent any particular entity—in computer parlance, the larger will be the "file."

The problem of "extent" is easier to grasp, since, if we have decided on the degree of resolution, then a hundred faces will take more memory in which to represent them than will one. Thus, as the playal expands the extent of its environment that can find its way into the learned-routine space, the amount of memory required will increase, since the question of the limiting resolution that can be achieved is settled, at least roughly, by the resolution present in the outer stream, as determined by the acuity of the associated senses.

Clearly, since the playal brain remained an ongoing site for the workings of natural selection, some way must have evolved for selecting the images that are stored in it, out of those that pour in on its outer stream, from the vast expanse of its environment. This leads to what must surely be one of the grandest questions that playals can pose to themselves: How do our brains select, from the vastness of everything around us, the small extent that is represented in our memories, at various degrees of resolution? Here, one can only surmise, as in the following sketch.

We begin with the description given above of the learned-routine space being gated *internally* by the initial challenge episodes originating in primal play. The implication there was that the gating is indiscriminate: the internal challenge episode gates the learned-routine space "on," and whatever happens to be incident on it is stored, until the internal challenge episode ends, and gates the learned-routine space "off." But this would be only the crudest form of gating, since one can imagine that it could be more selective, and begin to admit only those aspects of the environment that relate to the content of the gating episode itself, and not just its duration. In this way the playal would begin to store only those episodes "that are of interest to it," where "that are of interest to it" means "that relate directly to what it has already stored," and, evidently, copied and understood. In this way, what the playal has already stored becomes a selective filter on what else it will store!

Of course, such a filter would be imperfect, if only because of the nature of what is being filtered. But, precisely because of the imperfections, the playal would experience a slow increase in the extent of its environment that would be entertained in its memory. However, it would never cease to select what it entertains in its memory by filtering with what it has already selected, albeit somewhat imperfectly. In this way, the selective "interest" of the playal would serve as a limiter on the extent to which what it experiences would be committed to its memory. It would also limit how far outside its existing range of experience, at any time, the playal would extend its "curiosity." Such a process of filtering would also help to preserve the episodic structure of the learned-routine space, referred to in the previous section.

Another metaphor that we might apply comes from imagining that the nature of memory is such that it forms new "islands" only rarely, if ever. Instead, memory might proceed in a series of "peninsulas," in which case it would work as its own selective filter, limiting its expansion to additions to only what already exists in memory. Changing the metaphor again, it could be that memory is like a tree that never "takes up" the full space to which its branches are continually exposed, as it grows outward from only those places to which it had grown before. In this way, it displays its own identity as a tree, and, at least to some extent, a "selective memory" of what has befallen it during its own lifetime. But whether the process for limiting the content of memory is one of gating, or peninsulas, or branches, the formatting of memory-space, layer by layer in REM sleep, would remain a fixed part of its development, as would its episodic nature.

EFFECTS ON MAMMALS OF INCREASED NEED FOR BRAIN-SPACE

Even if playals can succeed, as apparently they can, in limiting the extent of their experience that is lodged in their memories, it remains the case that, in comparison to reptiles, they develop an enormous and increasing need for brain-space. Now, the physical nature and role of an increasingly elaborate playal brain is such that it needs, more and more critically, the protection of the tough, bony covering of the skull in which it is housed. In addition, there is some minimum fraction of what will become the adult brain with which the playal must be equipped at birth, in order to execute the functions then required of it. Thus, the bony structure that houses the playal brain at birth must occupy some minimum, but substantial fraction of its adult size.

Evidently, if practical procedures at birth are to continue to obtain particularly in a *mammal*, the skull in which its brain is contained cannot continue to grow indefinitely to accommodate the evolutionary pressure from copy-learned behavior; and so the components of the brain supporting less beneficial behavior, instinctive behavior in particular, might be expected to disappear and so free up at least some of the memory and processing space required by copy-learning.

Thus, any mutational tendencies to remove components of the brain that are associated with at least some instinctive programming can be expected to be supported by the process of natural selection. In this way, the mammal brain will be driven toward a state in which it will have lost some of its previously instinctive programming. It will have substituted for the availability, at birth, of a full instinctive program, a lengthening period of parental dependence and protection, during which it can equip itself with a beginning supply of artificial challenges and subsequent episodes, out of which it can build a repertoire of additional behavioral programs based on a relatively full version of its *own* past, individual experience.

In this respect the mammal has become a new kind of animal altogether, for it has come on a way of partially detaching itself from eons of evolutionary development of purely instinctive, genetically-determined behavior, replacing this, to the extent that the necessity for instinctive programming of its vital functions will allow, with a system of synthesized behavior that relies on real and artificial challenges arising in its own early and continuing life. Genetic

145

control has now been expanded to include the biology that goes to make unconscious episodes, pain markers, comparators of behavioral routines, massive memories in which new programs can be stored, and the like. In this way, the mammal is embarked on a whole new line of evolutionary development in which early dependence and fear based on its own past pain are integral parts.

It is worth noting that, if this is the case, then the present level of instinctive behavior in humans must represent a rough equilibrium condition arrived at in a retrograde evolutionary process that should entail occasional *fatal* consequences in the following way. Evidently, the pressure to occupy instinctive-program space must have stopped just short of our having to learn how to *breathe*, for instance, in early life, since this would surely have led to extinction. However, we do have quite a lot of copy-learned control over our breathing, as evidenced in singing, blowing bubbles and "holding our breath," for instance, and this suggests that such control as we do have is a reflection of the penetration of copy-learning into a previously purely instinctive system.

Since the point at which this penetration has ceased will always be subject to random genetic perturbations in the form of mutations, some of these perturbations would favor the further advance of copy-learning into instinctive space that drives breathing. But, even though the increment represented by such a mutation, by itself, might be quite minor, it could prove fatal to the young, since it would require more continuous, learned driving of breathing than any human, but especially an infant, could provide. And, in a moment, we will see what the related consequences of this can be for infants.

Evidently, such genetic perturbations would manifest themselves as "diseases," so we could call them "diseases of usurpation," to remind us that they are due to the continuing usurpation of instinctive function by copy-learning. It is, of course, precisely the fatal effects of such diseases of usurpation in the young that must limit the continuing effects of the evolutionary pressure of copy-learning on instinctive systems, since such diseases continually remove their carriers from the evolutionary stream before they have a chance to produce offspring. In a quite fundamental way, then, it is such diseases that, by their fatal effects, give practical expression to the *biological impossibility* of a creature whose brain would be fully programmed by copy-learning alone, having lost all its instinctive functions. (This, as you doubtless realize, is one of our greater blessings, since it relieves us of the necessity of understanding how we come to walk, or, even more mysteriously, how our liver, for instance, performs its marvels in a perpetually dark underworld!)

SUDDEN INFANT DEATH SYNDROME: SIDS

It seems certain that many of the so-called "sudden infant deaths"—as collected up in the Sudden Infant Death Syndrome, SIDS—which continue to resist explanation, are cases of such a disease of usurpation. As of now, there exists no substantial experimental confirmation of a genetic component to SIDS, beyond indications that siblings of previous victims of the disease are at measurably greater risk of being victims than the average infant. But, as I will now show, there are compelling reasons for believing that such deaths arise in this way: the fact that the infants die (a), *while sleeping*; (b), most frequently at *age about four months*; and (c), according to coroner's reports, *"from a lack of oxygen."*

And to see why this is so, begin by recalling that passages of REM sleep are characterized by extreme outpourings of signals from the *learned-routine space*, which lead to *paralysis* of *those muscles that respond to copy-learned routines in the waking state*. Of course, just after birth, an infant will have accumulated no copy-learned routines, but this will change, as the infant carries out copy learning in the early months of its normal development. Then, at the age of about four months (but certainly before twelve), the infant will have accumulated sufficient copy-learned routines *for these to take part fully in the processes involved in its REM sleep.*

And it is here that the risk of death makes its appearance. For, as, during REM sleep, the infant's growing copy-learned routines increase the paralysis they induce in *those muscles that respond to copy-learned routines in the waking state,* they eventually induce paralysis in those copy-learned routines that drive *breathing*. So, if the infant has *inherited just a little too much usurpation of its instinctive breathing,* then, when, in REM sleep, *its copy-learned routines will paralyze the muscles that are driven, in the waking state, by its copy-learned routines,* the remaining (un-usurped) *instinctive breathing* is insufficient to sustain adequate intake of oxygen. The final step in this process would involve the "sudden" death of an infant, aged about *four months, while sleeping,* apparently *"from a lack of oxygen."*

We can imagine from all this that the first minimal appearance of the control of behavior in the mammal by copy-learning set in motion a regenerative process of evolution in which a state of near equilibrium was reached eventually due, on the one hand, to the limit size of its head, prescribed by necessities at birth, and, on the other, to the limit to which space, previously occupied in its brain by programs for instinctive control, could be used up by copy-learning without impinging too much on the mammal's vital, instinctive functions.

In this way, increasing fractions of the behavior of the mammal came to be under the control of extremely elaborate copy-learned and socially-transmitted programs stored in its brain, compared and selected finally on the basis of its own past "success," as defined by pain. Thus, the behavior of the mammal had lost its exclusively instinctive, reptilian character, and even some of its semi-internal behavior, such as breathing, began to reflect the minimum practical limit of instinctive control.

INTERNAL CHALLENGE AND MADNESS

As we can see, the emergence of internal challenge in the mammal leads to a whole new range of possibilities that move it away from the limited behavioral confines of both its reptilian and early playal ancestors. But it would also seem certain that behavior derived from a process of internal challenge must, at least in some cases, lead ultimately to external behavior that is not appropriate to the survival of the playal in all the external circumstances in which it can find itself, since the power of the process is to be found precisely in the detachment of the associated learned behavior from the immediacy of challenges arising in external circumstances that relate to survival.

Of course, the more extreme forms of behavioral inappropriateness are referred to simply as "madness," and so we can see quite easily that one of the inescapable consequences of the emergence of behavior traceable to internal challenge in the playal will be its vulnerability to individual behavioral displays that will be classified as "mad." The details of this are well beyond anything I should be attempting here, but it is important to notice that, going along with the power flowing from such an "unconscious life," there was the certainty of occasional *individual* madness.

I stress "individual," because purely instinctive behavioral routines can also lead to "madness," but it will be general madness. This will occur when the environment that led to some particular form of behavior changes faster than the genetic process allows the instinctive behavior of a species to keep up with the change. When this happens, instinctive behavior will become every bit as mad as that mentioned earlier, but then *all* the members of the species affected will display the same "madness," and it will lack the individual character of that associated with copy-learning and internal challenge.

This kind of individual madness must already have been fully evident in the almost-human creatures that preceded us, but they would not have displayed the full range of human madness, since this, as the next chapter will show, must have involved a further major evolutionary development associated with storytelling.

SUMMARY AND EVIDENCE

Because the processes and developments that are being described are both numerous and complex, I have found it necessary, as you will have noticed, to proceed by a succession of long explanations which would tend, eventually, to get far out of step unless, from time to time, I collected them all up and put them back in step again as a new whole. I intend to do this now, and then proceed with some evidence, drawn from observations on the links between actual brains and actual behavior, which, I think you are going to agree, support the line of argument as it has been advanced to this point.

So far, I have traced the evolutionary specialization of the playal from the pre-playal into the playal stage. The creatures that characterize the pre-playal stage are those that do not play. These are made up of animals that do not eat other animals, as well as animals that do eat other animals, but which produce their young under such circumstances that they can escape the attacks of their parents, and so the play situation will never arise. These pre-playals, which include reptiles as a prominent class that is of particular interest, will be limited to the type of hereditary, genetically determined behavior that we refer to as "instinctive."

Of course, "instinctive" does not imply "simple." All it means is that behavioral routines are built up and linked together by biological processes that are determined by fixed arrangements that do not allow the animal to vary these routines significantly, based on what it experiences in its lifetime. Thus, the behavior of every such animal will be almost entirely dependent on where it is along the sequence of steps in its life that takes it toward death.

The playal orientation develops among the limited sub-set of the animals that eat other animals, and rear their young in close proximity to themselves. These will be animals that are still characterized by instinctive behavior, but with the addition of a new aspect which I have called "primal play," and which takes the form of a muted attack by parent on young. Both birds and mammals are either active, visible, "true" playals, or vestigial playals, having both evolved from reptiles via ancestors that were young-rearing animal-eaters. The vestigial playals are those that are now plant-eaters, and so have no reason, by virtue of their present diet, to play.

It is in the context of primal play, and the unique relation between parent and young which forms part of it, that the first trace of "copy-learning" develops in the brain of the playal, in that part of it that I have called the "learned-routine space." At the heart of copy-learning is the storage of actual nervous streams which allow repeating the behavior with which the nervous streams were originally associated. These nervous streams I have called "inner streams," and they are stored in the learned-routine space. This way of augmenting the playal's range of behavior no longer relies on repetition and success in applying instinctive routines. Instead, copy-learning achieves the copying of observed behavior by a process which, in principle, can lead to a newly acquired behavioral routine after a single observation.

Copy-learning will be present in both "true" and vestigial playals, among birds and mammals, since it grows out of primal play. Both "felt selfness" and "understanding" are simply integral parts of copy-learning, and so they come into the life of the playal, at this relatively early stage, with it. Copy-learning marks the shift from behavior based purely on genetically programmed routines to that which includes an expression of episodes in the life of the playal itself.

The emergence of the learned-routine space, and the associated copy-learned behavioral routines, led to a major change in the pattern of sleep in the playal, compared to what it had been in reptiles. In particular, the previously quiet sleep of reptiles is interrupted by passages that are accompanied by paralysis, and evidence of extreme stress. These passages, known as REM sleep, are found in mammals and birds, and serve to confirm the play-associated origins that they share.

A state of challenge pervades the playal during primal play, and hence during the emergence of copy-learning and the learned-routine space. This sets up the requirement that further access to the learned-routine space is only possible if the playal is in a state of challenge, either external or internal.

Another development takes place in a sub-set of the playals, and stems from the challenge situation present in primal play. This takes the form of "game-play," which needs signaling to initiate and sustain it. This, in turn, leads to the emergence of active-passive grammatical language, and the assignment of parts of the two hemispheres in the brain to the storing of the active and passive parts of the signals that language is constrained to have. The emergence of the active-passive structure of game-play and language is only possible in the presence of the "felt selfness" and "understanding" that form part of copy-learning.

A further development sees the emergence of a process for developing internally generated "unconscious" challenges, derived from episodes run in the brain of the playal, while it is awake. This leads to a set of new behaviors that can be copied into the learned-routine space, and which are derived from more general experiences in the life of the playal than those connected exclusively to external challenge, as in play. The acquisition of these new behaviors is accompanied, in mammals, by an increasing replacement, by copy-learned routines, of some of those programmed, instinctive behaviors that previously characterized the full external behavior of the playal. This is related to the large increase in the volume of memory necessitated by copy-learning generally, and the restriction on the size which a practical mammalian head can attain due to necessities at birth. The increasing usurpation of instinctive programs can lead to the death of an infant while sleeping, as the paralysis occurring in REM sleep shuts down the extreme copy-learned breathing in the infant. The process of internal challenge is inseparable from the possibility of attacks of individual madness.

This is the point at which we arrived near the end of the section that preceded this summary. What I shall do now is show that there is experimental evidence in support of the series of developments just summarized. Specifically, what will be looked for is evidence of the following:

1. Some part of the brain that can be associated with the pre-playal stage. This should be a relatively ancient and primitive part, devoted largely to instinctive control. It should be found in all reptiles, birds, and mammals.

2. Some other part of the brain that can be associated with the primal-play stage. This should be found in mammals and birds, but not in reptiles, and be identifiable as having evolved later than the primitive part in (1) above.

3. A third part of the brain that can be associated with the stage in the evolution of the playal at which game-play and active-passive language develop. It should also be related to a widened range of behavior which derives from internal challenge, and a reduction in instinctive control. In effect this part of the brain should suggest almost-human activity.

The evidence I'm seeking is summarized in the following passage from *The Brain* by Richard Restak:

> "If we remove the cerebral cortex – that part of our brain which has evolved over the past two million years or so – we essentially eliminate our humanity. Beneath the cortex is a brain that is not far different from that of a Bengal tiger, a French poodle, or an Arctic fox. We could, if we wish, remove even more to approximate the brain of a salamander or a rattlesnake.

> "Brain researcher Dr. Paul MacLean holds that we have three separate, intimately interconnected 'brains' that reflect our ancestral relationship to reptiles, early mammals and

late mammals. The first, the R-complex (the "R" refers to reptiles), is an expansion of the upper brainstem. Within this complex are the neural mechanisms responsible for behavior involved in self-preservation and preservation of the species. By experimenting with the R-complex in animals as far-reaching as squirrel monkeys, and turkeys, Paul MacLean has shown that it contains the programs responsible for hunting, homing, mating, establishing territory, and fighting. The second brain – the limbic system, which we share with all other mammals – deals with the emotional feelings that guide behavior. After destruction of part of the limbic system, young mammals '*cease to play and there are deficits in maternal behavior*', says Dr. MacLean. '*It is as though these animals regress toward a reptilian condition.*'

"The third 'brain', the cortex, is most highly developed in humans. It is a kind of problem-solving and memorizing device to aid the two older formations of the brain in the struggle for survival. Dr. MacLean compares the cortex to a 'computer' that can look into the future and anticipate the consequences of actions. The cerebral cortex furnishes us with our most human qualities: our language, our ability to reason, to deal with symbols and to develop a culture." (My emphases) (Restak, pages 137-8.)

It seems to me that the evidence from the work of Dr. MacLean is fully in support of the existence of the three parts of the brain that the foregoing play-based speculation leads one to expect. Particularly striking is the observation that, after destruction of part of the limbic system, the mammals involved "*cease to play.*" Evidently, the three parts of the brain that I was seeking can be identified as:

1. An expansion of the upper brainstem,

2. The limbic system,

3. The cortex.

The only comment that seems necessary relates to the cortex, which is associated with language, as required. However, as I shall show in the following chapter (and more evidence from Dr. Maclean's work will confirm), only a limited part of the cortex is associated with the final evolutionary step that made us human. But this does not alter in any significant way the force of the evidence provided, just now, from Dr. MacLean's remarkable work, which can be accessed via less popular sources. One of these, listed in the References, is the *Encyclopedia of Neuroscience*, where can be found Dr. MacLean's article "Triune Brain," which is as beautiful as it is brief.

CHAPTER VIII
THE HUMAN PLAYAL

THE SELF-SERVING STORYTELLER

As we have just been seeing, the internally-challenged playal begins to display more of those features that we think of as human. However, what happens in the playal to render it truly human? This is the question that I now begin to answer.

The evolution of the playal has, so far, led to the emergence in it of two major lines of development. One of these has led to the ability to re-run, inside the playal's brain, sequences of images originating in the episodes and challenges of primal play; and when the playal begins to challenge *itself* in this way, the limited group of episodes originating in primal play is augmented by the presence, in its learned-routine space, of more general episodic streams not having any particular connection to *external* challenge as this exists in primal play. The other line of development has led to the ability to construct and interpret isolated, spoken grammatical sentence-pairs, originating in the structure of the image-pairs that accompany game-play.

But this still leaves a huge gap between such a playal and you-and-me, in that I can tell you of all sorts of things that I saw and that I did and even that I dreamt. I can tell you of all sorts of things from my past, and the pasts of others, which can then fill a part of *your* present and ultimately *your* past. All that our playal can do, so far, is signal in single sentences about its present, while I can deliver long strings of sentences connected up so as to describe long sequences of images that resemble the image-sequences both outside and inside my head, from the past. And so we come on the essentials of what our playal had to do to move toward being human.

What it had to do is move in the direction of connecting up its two major lines of development; that is, the playal must evolve in such a way as to join up its ability to make image-sequences, on the one hand, with its ability to

155

make single sentences, on the other, so as to yield a new composite ability to use strings of sentences to describe and recount the goings-on in the images both inside and outside its head.

It would hardly seem that this could be "all" the playal would have to do to close the remaining gap and become human, but this is only because, being so fully developed as effortless storytellers, we easily overlook the remarkable processes that must underlie the evolutionary emergence of the spoken story.

To see what some of these might be, let us begin by noticing that there is virtually no survival value attached to being able to recite one's own experiences to oneself, since one would be reciting what one already knows. So any value that might come from the ability to recount past experiences in spoken strings of sentences must come from one's ability to make use of the stories recited by *others*. This is the first clue to the emergence of narration: For narration to have any evolutionary significance, the playal that is going to acquire the power to narrate must, at the same time, acquire the ability to receive the narrations of others *and make use of them in ways that affect survival,* that is, in ways that affect external behavior. For, in evolutionary terms, it does *you* no good at all, in fact, it's downright harmful, to load up *your* brain with volumes of *my* stories unless *you* can find a way of using them, sooner or later, to change *your* behavior so as to improve *your* chances of survival.

And so the very first thing that some emerging, almost-human playal had to begin embodying, as a part of the capacity to narrate, was the capacity to transform *received* narration, inside its brain, into images that can help to meet its *own* needs. What this implies is that the playal must be able to *copy the behavior of some actor that is described in the narration.* Without this ability to copy the behavior of such an actor, narrated episodes would have remained useless encumbrances in the brain of a playal headed for extinction.

NARRATION AND COPY-LEARNING

Copying the behavior of an actor described in narration is not the same as copying the behavior of another playal encountered face to face, and it is important to see why. Beginning with a review, we recall that, prior to the emergence of narration, a creature whose behavior was copied formed part of the *direct* experience of the playal, and, in addition to exhibiting the behavior to be copied, it had concrete features which helped to define it as a real, individual creature. When copy-learning occurred in these circumstances, the copier extracted the behavior of the copied creature, free of its concrete features. Thus, the process of copy-learning from a real creature always entails the process of abstraction, which consists of stripping the copied creature of its real, concrete features, and leaving the copier with an element of "pure" behavior that then becomes part of its own behavioral program.

In contrast to a real creature that is present, and whose behavior can be sensed *directly* and matched, the actor described in narration is *absent*, and is represented by a set of what amounts to instructions only about how to make the actor appear, and make it display behavior that could be copied. Thus, before the playal's copy-learning processes can be applied, some step must occur that will make the absent actor appear, and display behavior that can be copied—the ghost in the narrative must be transformed into a "something," a "creature" having *behavior* that can be copied. Evidently, it is the listener-copier, not the narrator, that must perform this step, and it is here that we come upon the truly new and remarkable aspect of narration.

What we can see, then, is that the listener-copier will have to perform essentially two operations: the now old operation that copies behavior from a real actor, and a new one that transforms the "actor" in the narrative into a "creature" having *behavior* that can be copied. And if this new operation is called "animation," then narration can be said to consist of two operations: animation and copying, *in that order*. And, given this order, narration becomes linked to *animation* as an absolutely unavoidable concomitant.

Since *all* the playals that learn to narrate must be able to perform this feat of animation, we can see that narration must imply a relationship between how we narrate and how we process narration in our brains, in order that the stories we narrate might be transformable into episodes into which others can "insert" *themselves,* so as to effect animation. But for these reciprocal properties, narration would have remained a trivial option that would never have been able to slide itself into the stream of evolution.

It is clear from the foregoing that the acquisition of narration, with its unavoidable animation, must have come at the cost of considerable modification in the brain of the playal. So, what benefits could secure for such modification the support of natural selection?

THE BENEFITS OF STORYTELLING

Fortunately, the benefits of storytelling are far easier to explain than the processes that support it, since what it confers on a playal is the capacity to pour into the brain of each member of its species, by word of mouth, significant portions of the experience of all the other members with whom it comes in contact, and even other members beyond that. Consequently, not only does the behavioral repertoire of each pre-human, as explained before, show its release from instinct as it is enriched by its own increasingly complex experience, but now, by means of narration, this repertoire is enriched even further by the shared experience of these new storytelling playals that begin to appear in increasing numbers.

Prior to the emergence of narration, the behavior of each playal was limited to the exploitation of second-hand experience, of mixed vintage, stored in its genes, augmented by its *own* first-hand experience gathered in copy-learning. However, narration provided the playal with a multitude of *new* sources of second-hand experience, all now of *recent* vintage, on which it can draw for tuning its behavior to its *current* environment. Also, except for the mixing that occurs in mating, the experience stored in genes is loaded into the playal in *series*, generation after generation. But, beginning with copy-learning, and increasing vastly with narration, experience is loaded into the playal in *parallel*, from like creatures around it, and even those beyond.

Narration must, therefore, have given rise to explosive benefits for our ancestors, so much so that one wonders whether the rate at which the rest of our biology could evolve has ever succeeded in coping with the enormously increased rate of intake of experience that narration made possible. This becomes an even more pressing wonder when we realize that animation would not have been limited to the humanoid actors alone in narration, as I shall explain a little later on.

NARRATION AND HUMANNESS

As we can see, the emergence of narration was accompanied by a major transformation in the relation of the playal to its environment. I believe that this was the final transformation during which the human species emerged. In order to lend more credence to this position, I shall discuss a number of derivatives of narration that have clear expression in humans. The derivatives are presented in the context of narration and its relation to the following items:

Grammar;

Selfness;

Being Led;

Pain;

Non-human Entities;

Myths;

Empathy;

Abstraction;

Untruth;

Unreality;

Schizophrenia.

Schizophrenia is included in the list because, as I shall show, it is almost certainly an illness associated with failures, of one kind or another, in the biological system that supports narration—indeed, that supports our final humanness. In order to facilitate the demonstration of this, the discussion of each of the items that precede it will include a sketch of what would be expected to be the effect on behavior of a failure in the biological system supporting narration.

Narration and Grammar

When the behavior of a real creature was being copied in the earliest from of copy-learning, the copying playal was working with itself and directly with a second playal only. Thus, in terms of grammar, the playal was working in the first and second "persons" exclusively, that is, under conditions in which English makes use of "I" and "You (Thou)." However, in the case of narration, the second person is the narrator, and so the absent actor referred to and described in the narrative becomes a *third* person, that is, the "person" who, in English, is referred to by "he" or "she" (him, her, it), or by a "noun" such as "tree," or "river," or "sun," or Robert, or Louise, or Carol.

We can see, then, that the emergence of narration involves not only the introduction into grammar of the third person and nouns, but, equally remarkably, of a process in which the playal transmutes a third "person," that is absent, into a second "person" which "exists" within the playal itself, in order that copy-learning, with its abstraction of "pure" behavior, can transform this second "person" into a part of the playal itself, running in the first "person."

Going to a different aspect of grammar, you will recall that, in Chapter VI, I drew attention to the absence in discourse, as we make use of it toady, of any apparent necessity to employ the *passive* form of the sentence, and the need to explain this, in view of the crucial role played by the paired, active-passive form in the emergence of grammar and language. We are now in a position to see what that explanation might be.

Begin by recalling further that the active-passive form of the sentence is an integral part of the I-YOU, first- and second-person situation of game-play and pre-narration language. However, the first- and third-person context of narration changes the relation from that between an "I" and a real "YOU," with roles that can be interchanged completely as in game-play, to one between an "I" and a purely artificial "you," for which the active-passive form no longer has any behavioral significance, since there is no symmetrical way of playing a game with an *absent* actor.

Thus, while the active-passive pair had been, in the earliest phase of the emergence of grammar and language, an integral part of the game-play situation in which it arose, such pairing ceases to have any basis in the later case of narration; and so, in the structure of this later form of grammar and language, the passive half of what started out as a sentence-pair is reduced to a purely vestigial trace of its previously essential role. As a consequence, in narration, the passive "voice" becomes just a stylistic convenience that makes use of a structural form whose origin and significance do not reside in narration itself. And since narration has come to be such a large part of the context in which language is employed, the passive voice appears to be a sort of curious appendage to language generally. And that is not just how "all the foregoing strikes me," but also how "I am struck by all the foregoing"!

This suggests that any attempt, such as that made by Chomsky, to find a "universal grammar" that underlies the whole of language, as it presently exists, is not likely to extend easily into narration, since narration has lost a large part of an essential transformational component. But it is clear that, just as Chomsky insisted, some kind of universal, genetically determined active-passive grammar *did* form the basis of language. What remains of this, that is of significance for narration, is the link between the *structure* of the sentence and its *meaning*, but this is fully expressed in what are really the single-sentence half-signals of narration, without any need for the passive and its transformations, in order to achieve the kind of copy-learning and understanding on which the survival value of narration depends. In effect, narration substitutes for the power of the externally based, direct, active-passive relation of game-play, that of the internally-based, indirect, active-only relations that depend on animation.

It also appears to be the case that, as shown in the experiments with the children mentioned in Chapter VI, who had undergone the removal of one or the other hemisphere of their brains, passive sentences are transformed, still now, as they must have been at the time of the earliest emergence of language, by a process that needs both hemispheres for proper execution. This is probably why passive sentences tend to be associated with a certain feeling of clumsiness—witness how "I am struck by all the foregoing."

This also allows us to see why language, as it now exists, appears to be associated mainly with one hemisphere, since the separation and symmetry of two hemispheres, required for active-passive grammar, is no longer required for narration, which accounts for much of the present content of language. However, it would seem reasonable that language which currently employs the first and second person only, that is, language associated with the reality of direct experience, should still display the kind of sharing of hemispheres that must have formed the basis of the earliest spoken language that preceded narration.

We can summarize much of this by imagining an evolutionary sequence that has instinctive, primal play as the first level of the artificial, having a "grammar" limited to the first person. This is followed by game-play as a second level of the artificial, having an active-passive grammar that includes a first and a second person. Finally, narration provides a third level of the artificial, having a grammar that includes a first, second and absent third person, including nouns. While the third-person part of this grammar is no longer actively supported by an active-passive structure, the sentences that make up the related parts of the language still have a structure that is determined by the original active-passive patterns.

Such an evolutionary sequence for grammar helps us imagine the way in which the actual process of narration could have emerged. Beginning with game-play between two playals, the first and second person would have sufficed for the necessary communication. As game-play expanded, the behavior of more and more items in the environment in which it proceeds would form part of what could be communicated with benefit. Such expansion could begin with nothing more complicated than game-play between *more than two siblings*. But as soon as the items involved moved beyond the immediate I-YOU context of game-play between two playals, the inclusion of such items would make appeal to new gestures, of which *pointing* would almost certainly be the most useful and prominent. In this way, a third "person" would become part of what is being communicated, and find a *gestural* basis for its inclusion in language.

But even though such an increasingly distant third "person" could still be experienced directly, this would no longer be in the immediate I-YOU context that is characteristic of game-play between two playals. Thus, there begins a slow drifting of the subjects of language from immediate, to more and more distant actors, until pointing itself fails. In the final stage of this slow process, the third "person" becomes an entirely absent phantom, no longer addressable even by pointing, indeed, a phantom inaccessible to any direct sensing, and which can be conveyed only in the full indirectness of surrogates for senses assembled from strings of sentences. The ultimate form of the actor in narration becomes that of such a phantom, totally without any direct existence so far as listeners are concerned, and able to affect their survival only to the extent that each of them can endow it with some sort of "copyable behavior" by animation.

It seems reasonable to suppose that a failure in the biological system that supports animation could leave the victim unable to change his or her own behavior based on that of the actors described in narration, that is, the actors described in narration would have no "meaning." It could also limit the victim to the use of the first and second person only, thus producing the effect of a substantial loss of speech.

NARRATION AND SELFNESS

As explained in Chapter III, the earliest playals that were capable of copy-learning had acquired felt selfness. It is this selfness that allows any particular playal, in the presence of behavior acquired from other playals by copy-learning, to continue to "feel" itself as the source of the programs of behavior that determine its relations with its environment. Thus, selfness contributes to the "feeling" of being controlled from "inside," rather than from outside. All of this relates to copy-learning performed on other real playals, as a part of direct experience.

In the case of copy-learning carried out on the phantom actors in narration, the new processes are more complex, and although they must also lead normally to the retention of the feeling of selfness in the playal, they have new ways of failing that are associated with the new processes themselves. One such would be that in which a failure of the abstraction process occurs either because of a defect in the way that animation is performed, or a defect in the way in which an animated actor is presented to the processes of copy-learning.

Thus, we can expect particular cases of illness linked to discrepancies in "selfness," in which the victim will report a loss of control from "inside," with the substitution of control that comes from "outside," but particularly from "outside actors" originating in *narration*. As we will see, these "outside actors" can take a variety of forms.

NARRATION AND BEING LED

There are some special cases of narration that can be grouped around the situation in which the narrator recites a story in which *he or she* is an actor. In one such case, the narrator recites a story about the narrator alone as actor, and, because of the nature of copy-learning, lays the groundwork for episodes in which the behavior of the listener will reflect directly the narrated behavior of the *narrator*. In such a case, the narrator could be said to be able to "lead" the listener, that is, to function as a "leader" and exercise some degree of "control" over the behavior of the listener. This can be conveniently referred to as the "hypnotic" case of narration.

The limiting version of the hypnotic case is that in which the narrator is also the listener, that is, recites to him- or herself about him- or herself alone. In such a situation, and again because of the nature of copy-learning, the narrator-listener would be able to achieve a degree of regenerative auto-control of behavior, that is, of emphasis in behavior of certain previously stored image-sequences, by narration; and this emphasis could be further emphasized by further narration, and so on. By extension, this should be referred to as the "self-hypnotic" case.

Evidently, we have here come on the fact that the emergence of narration is inseparable from the emergence of the ability of one playal to control the behavior of another by means of storytelling, and "suggestion." Thus, the narrating playal is, of necessity, the "suggestible" playal.

A failure in the biological system supporting narration would render the individual so affected unable to follow suggestions as these would arise in narration, and the person would appear to be stubborn and incorrigible. It would also be difficult or impossible to induce hypnosis in such a person.

NARRATION AND PAIN

It is necessary to pain-label the narration-derived behavioral routines so as to have some scale on which to compare them for possible use. So it is important to try to imagine, especially for some later parts of what I am proposing, how behavior, which is reconstructed by the listener from the word-streams in narration, comes to be pain-labeled.

Evidently, the listener can achieve a substantial benefit from narration if the degree of "success" of the actors is registered in the *listener's* pain, and preserved with the reconstructed sequence in the brain of the listener, since, in this way, by drawing on the experience of the actors in the narration, the listener can avoid unnecessary risk of failure when the learned sequence is first used.

Thus, not only must the process of animation transform the pain-related words associated with an actor into attached parts of images, but the process of abstraction must not strip away this part of the "concrete" aspects of the actor. This could reasonably be expected to become the basis of the normal process, but it is worth noting that a failure in this process could lead to a playal with behavioral routines that were derived from narration, and which have lost all reference to pain, that is, to the pain of the actors from which they were copied. A playal so affected would, as this relates to behavior derived in this way, lose some of what we call its "fear," and could be expected to display "reckless" behavior, which exposes it to greater than "normal" danger.

A particularly interesting case of such behavior would be that in which the behavior learned from narration relates to ways of taking the playal's own life. In such cases, the first use of the acquired behavioral routine would often be final, since the playal would, in cases of "successful" application of what had been learned, have no opportunity to re-label the behavioral routine based on its *own* experience of pain, and so avoid such behavior in subsequent situations. Evidently, this would serve to perpetuate the spread, by narration, of a behavioral routine that ensures its suicidal deadliness precisely because it is never possible to assign it a pain label based on any actual experience.

Thus, a failure in the biology of the narration system can be expected to lead to behavior that contains more than the normal level of exposure to danger, including a higher than normal rate of attempts at suicide.

Narration and Non-Human Entities

Although the process of copy-learning coupled to animation must have found its earliest application to humanoid actors portrayed in narration, it need not be confined to them, and can, evidently, be applied to non-human animals, trees, mountains, rocks, rivers, indeed, anything whatever that could become an "actor" in the narrative of some storyteller. However, while the benefits in the case of humanoid actors are easily identified, those associated with other animals, for instance, might seem less so, at least when the process first emerges. But, given the clear benefits associated with the process when applied to humanoid actors, it is not difficult to imagine that its application would begin in this way, but spread to other actors from which the benefits, although more difficult to trace, are still real.

To see just how real, all we need do is notice that the animation of any entity to be found in narration leads directly to its "behavior" becoming a candidate for copy-learning, and hence for *understanding*. This is of enormous significance, since we can see here the way in which narration and animation finally open up the entire universe of the playal, about which stories can be told, to its ability to copy-learn, that is, to its ability to *understand*. What the power of narration allows us to do is assign a "behavior" to any entity at all that can form part of a story, and, in assigning such behavior, we become able to understand it in terms of that behavior, because that is what we can copy. But an essential part of being able to assign such behavior, and ultimately being able to copy and understand, is the necessity to "animate" the entity, and, in a curious way, be able to "behave like it."

We can also see how we can come to understand entities with which we have no direct experience, and about which we can only tell stories, that is, entities that don't exist at all, except in the form of an "actor" in a story. This also begins to expose the origin of the links between apparently separate activities such as the production of metaphors, stories, and theories, as well as the link between animism and the way in which we understand the world generally. But these are better left to later.

In order to represent inanimate actors in narration, the playal would have had to begin to use "common" nouns, in addition to the pronouns and proper nouns that it began by using to represent absent humanoid creatures. Nouns can therefore be viewed as the vehicles in which the distant, inanimate actors in narration ride into the brain of the playal; they were essential to the extension of narration to include all the entities that are now to be found in the stories told by humans.

This, as you will have realized, ties up a loose end, left dangling in Chapter III, having to do with the significance of copy-learning the behavior of inanimate objects. Now we can see not only how this comes about, but also the extent to which a whole array of what we regard as truly human capacities are linked to this aspect of copy-learning and its expression in the remarkable processes that sustain the behavior of the storyteller.

Here we come on a whole host of ways in which behavior could become disturbed by a failure in the biology of narration. A few of these are as follows:

1. Inability to construct whole stories from long strings of sentences;

2. Inability to understand entities and behaviors other than those originating in direct experience, and hence an inability to work with proverbs, metaphors, theories and other types of entities arising in narration only; inability to work with the "abstract"; limitation to the "concrete";

3. Inability to make use of the notion of the "behavior" of inanimate entities.

NARRATION AND MYTHS

It will be convenient to distinguish here between a "phenomenon" and the "entities" just discussed. The easiest way to explain the distinction is with an example: by "entity" is meant something like "sun," and, by "phenomenon," something like "sunrise"; that is, "entity" refers to a more or less separable item in the environment, and "phenomenon" to an episodic sequence in it. Clearly, this distinction is one of convenience in explanation only, since we can find a phenomenon within a phenomenon, and then the "inner" phenomenon will behave like what I have called an "entity."

Now, as the benefits of narration are expressed through the behavior of the playal and the filtering of natural selection, the possibility arises of copy-learning the behavior of a phenomenon, as distinct from that of its simpler entities. A way of accomplishing this, which is within the behavioral range of the playal that has begun the process of animation and narration, is for the narration *itself* to include the animation of the phenomenon, and so reduce the process of copy-learning to that of copy-learning the behavior of the actor in the narration that animates the phenomenon. Thus, in the case of sunrise, the narration itself can contain the story of an "actor" that rises early and begins to drive a flaming chariot across the sky. This becomes the behavior that is copied in the brain of the playal, that is, it becomes the "understanding" of the phenomenon.

Evidently, the version of the phenomenon that is provided in the narration is what we call a "myth," and the driver of the chariot is a "mythological character." What this demonstrates is the essential role that myths play, in the narrating brain, in its arriving at the understanding of phenomena, and the extent to which mythological characters provide the actors on which copy-learning can operate. And so we come quite directly on the "gods" of mythology, which represent simply the actors necessary to animate phenomena of major extent, the ultimate god being the mythological character necessary to portray the behavior of the environment as a whole, and so bring this ultimate "phenomenon" within the framework of copy-learning, that is, of understanding.

It is important to see the absolutely essential role that myths play in the evolution of the understanding that characterizes the narrating brain, since there is a tendency to treat myths as somehow not based on anything "real," and so belonging to a trivial part of the fringes of human development and existence, of interest only to the most specialized of anthropologists. This is reinforced by the belief that, somehow or other, "science" has succeeded in providing humanity with a way of *escaping* this mythological basis for understanding the behavior of the phenomena that pervade the environment. But, as I shall show in considerable detail in Chapter IX, this is certainly not the case. For, so far as the mythological aspect of its development is concerned, all that science does is replace one kind of distinctly animated myth, in which the actors resemble players with which we are "familiar," and whose behavior we copy-learn as an everyday part of human development, with another kind of myth which is limited to a single major actor-god known as "nature," endowed with the power of "laws," and which, in the narrations that constitute science, replaces the older overall, explicitly "humanoid," actor-god.

If this necessity for the existence of mythological characters, including gods, is combined with the previously explained potential for loss of selfness linked to malfunctioning of the narration processes of a brain, we can see that an illness related to such loss of selfness would include not just reports of control of behavior from "outside," but particularly control by mythological characters originating in narration, including control by *gods*.

It is important to distinguish between the evident necessity for the normal narrating brain to entertain the presence, within it, of mythical characters, and the non-normal and potentially self-destructive "feeling" in such a brain of *outside* control effected by such characters, and which represents a breakdown of "felt" selfness. We should also notice, by recalling the connection between narration and suggestibility, that it should be possible to instill in the normal narrating brain, precisely because it is normal, the "feeling" that it is controlled by mythical characters, but this control would then be felt as coming from *inside* the brain, and fall within the normal control of "felt" selfness. Indeed, the brain affected in this way would be said to have been "taught" to understand the portion of its environment represented in the myth acquired in this way, and would be viewed as "holding one of the normal views of the way the environment works." (The question of whether a particular way of understanding the environment, based on a certain group of myths, increases the chances of survival is treated in Chapter IX, in a discussion of the nature of science.)

A picture of the place of myths in a narrating brain allows us to situate within it what we know as a "mystery," since a "mystery" then becomes simply a phenomenon with which it is not possible for such a brain to associate any form of even mythical actors or narration. Thus, a mystery lies outside the boundary of understanding associated with a particular brain, and it can only be brought inside this boundary by behavior that we refer to as "discovery," that is, by behavior that provides a myth that matches, at some level or other, the behavior embedded in the mystery. Evidently, the myth must include such actors as might be necessary to support a narration, and is, in a significant sense, simply a copy of the phenomenon that constituted the mystery. Another way of viewing a mystery is therefore as a phenomenon that it is not possible to map into a particular narrating brain, even by using all the capacity for myth-making and narration that it *then* possesses. Thus, a discovery will generally originate in a single brain as a local "mutation," and need to be taught to other brains, affected by the same mystery, by a process of narration and copy-learning.

We can see from this that a failure in the biological system supporting narration could manifest itself in the incapacity to penetrate mystery by the fashioning of myths that "explain" experience. In the most extreme case, this would severely limit the capacity of narration to support coy-learning, and reduce copied behavior to the simple imitation of the directly observable behavior of other creatures, including, for instance, their stance, walk and vocalization; for this would then constitute the limit of understanding of them, that is, the limit beyond which all is mystery, as would have been the case for pre-narration creatures.

NARRATION AND EMPATHY

If, as I am asserting, storytelling constituted the last humanizing step in the evolution of the playal, then it should be possible to find in this step the basis of the human feature that we identify as "empathy," since none of the evolutionary developments identified as occurring previously provide such a basis.

In the narrowest view that we can have of empathy, it is manifested in the capacity of one human to "identify with" the experiences of another; it is the basis of the ability to "feel for someone else." An example of this would flow from a question such as: How could an individual pre-narration humanoid creature come to know that dreaming, for instance, was not some kind of unique personal experience, but was experienced by others? Evidently, the answer is that such a creature could not come to know this, with all the fantastic behavioral ramifications that such an answer must imply. Indeed, only the later-developing capacity to narrate could serve to transform the hidden, individual experience of dreaming into a "shared experience," and so bring the derived waking behavior within the framework of understanding.

But in a broader sense, empathy is manifested in the capacity to identify with the "experiences" of other more general animate and inanimate entities than just humans. Is there anything specific about narration that can be associated with the development of such a broadly based empathy? The answer is clearly yes, since the whole appearance of empathy in the human can be linked directly to the process of animation, which is an essential concomitant of narration, and to the internalization of the actors in narration that are a necessary part of it.

We can see some of the consequences of this by noticing that in the earlier playals who preceded us, the act of offensive killing was linked to hunger. But we can observe quite easily in ourselves the fact that hunger is no longer linked instinctively to killing, and so the power of offensive killing which was, in earlier playals, under instinctive control, is, in us, under the control of copy-learning. Thus, at least some of this control must be related to the copy-learning associated with narration, and it is not difficult to see that much of this must come from animation and its expression in empathy.

Evidently, such restraint on killing must come from the internalized pain of the actors in narration which induce in us some version of the pain, as narration conveys it, that would accompany their being killed. But all of this must depend so critically on little details of the processes of animation, that it would take just some minor failure in them to remove all control from the human power to kill, and lead to an illness in those so afflicted that would transform them into sources of great danger to the other creatures nearby. Although the case is made in the context of killing, it could be made in the context of cruelty more generally.

We can therefore expect to find, resulting from a failure in the portion of the brain supporting narration, an illness that is related to the breakdown of empathy, in which the human playal will display uncontrolled acts of cruelty and killing.

Since empathy tends to be linked to being "good," it should be noted that its emergence in the playal is simply a consequence of the necessities and the self-serving advantages associated with being able to apply copy-learning in narration. In particular, empathy does not signal the beginning of a new non-evolutionary era marked by altruism and the denial of self-interest, and it is important to see that this is so, if we are to demonstrate the ability of a consistent evolutionary argument to lead us from reptiles to ourselves.

It is also important in order to ward off the persistent tendency to avoid the harshness of evolutionary explanation which, the more we succeed with it, the more it tends to reveal, in a distinctly non-traditional light, those of our features that we are accustomed to seeing as a special kind of gratuitous human "goodness," born in some non-evolutionary space, that was not accessible to earlier creatures.

But interestingly enough, far from denying human goodness, the foregoing argument tends to show it simply as a biologically necessary rather than gratuitous part of *normalcy* in us. What the argument does deny is the basis of the vanity inherent in *noblesse oblige*. It would seem that normal humans have to be "good," evolution has made it that way, and we have no choice in the matter. Whether this will be a major factor leading to our ultimate extinction is an engaging but entirely different evolutionary question.

NARRATION AND ABSTRACTION

In the original form of copy-learning, as it occurred in primal play, the playal succeeded in copying "pure" behavior from a real, external parent, with one degree of accuracy or another. In the case of a failure in the process, the accuracy simply deteriorated, and the playal needed more and more attempts to achieve a certain accuracy, until finally, in the worst case, it would require so many attempts that it would achieve no copying of behavior at all.

However, when the copy-learning process is coupled to the animation that must precede it in narration, the situation changes, since the playal is then no longer working with a real, external creature, but with an internalized version of an actor, reconstituted from narration. Thus, the entity presented for copying is drastically altered, and the process of copying does not now simply end up with either an accurate version of "pure" behavior if it is proceeding normally, or an inaccurate version if it fails, as in the original case of primal play; rather, in the presence of a failure, it can now produce "contaminated" behavior that carries different levels of vestiges of the internalized actor.

When such a failure is present, the playal can display not just "pure" behavior, of different degrees of accuracy, learned from the actor, but some of the internalized actor's other features as well, as if they were its own genuine features; the failure does, in fact, lead to a partial loss of the felt selfness of the playal. In the extreme case of this, the playal simply presents itself as if it *were* the actor, because this is the way it "feels," and its behavior would then be said to show signs of "delusion."

Thus, a failure in the biology supporting narration can lead to behavior characterized by delusion involving actors derived from narration. Of course, normal people display singular selfness, so delusion is very striking, and tends to dominate what are taken to be the indicators of such failure.

NARRATION AND UNTRUTH

Since a narration can refer to an entity ("actor" in the most general sense, and its behavior) that does not have any existence outside the narration itself, the possibility exists that a narration could present just such an entity, which *does not* have any existence outside the narration, *as if* it had such outside existence. When a narration contains such a combination of "does not" and "as if," it is said to contain an "untruth" or "falsehood." A narration that does not contain an untruth or falsehood is said to be "true."

Thus, a failure in the biological system supporting narration should lead, oddly enough, to a *reduction* in the production of untruths, simply because the failure leads to a reduction in the ability to construct a whole narrative that can convey untruth. In effect, what the failure will lead to would be better called "confused-truth," rather than "un-truth." (This raises the interesting question as to whether delusions constitute untruths.)

It is interesting to notice that the ability to generate narrations containing untruths is closely related to the ability to indulge in "make-believe," since, to get from the narration containing untruth to make-believe, all we need do is add to the narration containing untruth an indication that it *does* contain a (presumably "beneficial") untruth. Such indications take a variety of forms, all the way from the introductory "Once upon a time" of the spinner of fairy-tales, to the "Let us assume (make-believe) that" of the mathematician. Evidently, a failure in the biology supporting narration could lead to an inability to indulge in make-believe, of which one manifestation would be an inability to be involved in the fun of a Grimm or a Pythagoras.

NARRATION AND UNREALITY

Prior to the emergence of narration, the outer stream of the playal carried a version of its own, actual experience. All the actors in these outer streams had actually been encountered by the playal. However, that situation changes with narration, since all the actors conveyed in narration are simply phantoms that must be reconstituted by the playal before it can make beneficial use of them by copying, and some of them can even be the bases of untruths. Thus, we can distinguish between the actors present in the pre- and post-narration brain by saying that, while those in the earlier brain are all "real" actors, those in the later brain include animated and mythical actors, which can be called "unreal."

It seems reasonable, and I shall give evidence in support of this later, that, as narration evolved, the biological components supporting it occupied a separate, identifiable segment of the brain; that is, as narration evolved, a barrier between the real and the unreal must have evolved with it. This would mean that there would be one space in the brain for real actors, and another for the actors of narration. However, if the biology supporting narration should fail in such a way that the barrier became "leaky," then actors from one of the spaces could migrate into that normally occupied by the others, and mix with them, so that the distinction between real and unreal actors, which comes easily from normally separate "addresses" in the one space or the other, would be lost. In this way, we can see that a failure in the biology supporting the narration system could lead to the inability to distinguish between the real and the unreal.

We can extend this into a most interesting area by noticing that, since the grammar associated with the *actors* in narration is confined to the *third person*, if the animation of the actors includes their power to speak, the grammar of their speech would be confined to the use of the third person. In addition, as a result of animation, all the animated actors, residing in the brain of a particular person "belong to" the *same* person, namely the person whose brain animated and hosts all the actors. Thus, if the actors were to speak to and about one another, they would speak about the *host*, all in the third person. This would not, in itself, be unusual or disturbing to the host, so long as the actors were originating in the space normally assigned to narration, and so were clearly distinguishable as unreal. However, in the presence of the kind of failure just mentioned, in which the host becomes unable to distinguish between the real and the unreal, the host will hear "real" actors speaking about him- or her-self, in the *third person*. And among these "real" actors will, on occasion, be found the gods of myths.

NARRATION AND SCHIZOPHRENIA

Given the considerable biological complexities that must underlie the development of narration and its related processes, we can safely assume, and I shall confirm this with experimental evidence later, that there exists some limited, identifiable segment of the human brain that was responsible for this development. I shall refer to this segment of the brain as the "narration complex," and assume that it is responsible for the functions associated with animation as well as the links between animated actors and the older parts of the brain connected with copy-learning and the associated abstraction.

Thus, the narration complex will be responsible for a long list of human phenomena and functions of the type discussed above. These include narration itself and the third-person components of grammar; the whole set of relations connected with empathy, which include the capacity to "feel for others"; the pain-labeling of the copied behavior of the actors in stories; the phenomenon of suggestibility and the possibility of being led and hypnotized by means of narration; the ability to animate and copy the behavior of non-human entities, and so understand and construct theories about them, as well as the ability to understand the behavior of entities that exist only in narrative form; the ability to construct untruths; the presence in the brain of the unreal, and its separateness from the real.

All these, almost startling in their humanness and range, come with the narration complex when it functions normally, as mutations and natural selection have determined "normal" function. But, what mutations can be a part of doing, they can also be a part of undoing, and it is the behavior that might be expected from such undoing that, in my view, constitutes the complex departures from the normal that are witnessed in the illnesses that go to make up *schizophrenia.* In support of this view, I shall quote some of the more prominent of the numerous observed symptoms of schizophrenia, so that they can be compared with the behavior expected from failures of the narration complex discussed earlier.

But before listing some of the symptoms, I should point out that, without an evolutionary model of the origin of schizophrenia, the genetic nature and origin of the illness is not easy to be convinced of. Thus, although the foregoing discussion would lead, as a completely natural consequence, to the conclusion that any mutations that would affect the development of the narration complex would lead to schizophrenia, with at least some of its possible symptoms, as a *hereditary* illness, you will notice, in the descriptions of the symptoms to be quoted, a slight tendency to "question" whether the illness could have a genetic origin.

The symptoms of schizophrenia have been studied extensively, and described by many people, in many different ways. I have selected passages from two of these descriptions, and the first selection, from *Brain, Mind and Behavior*, by Bloom, Lazerson and Hofstadter, in which all the emphases have been added by me, is as follows:

"Converging lines of research have led to the modern view that schizophrenic diseases have a *biological* basis . . . However, before we discuss this evidence in detail, it is proper to point out that no specific causes of schizophrenia have yet been directly identified." (Page 265)

"The genetics of schizophrenia are relatively complicated, but still speak a rather clear message. Some *inheritable* predisposing 'factor' can lead to the development of schizophrenia." (Page 267)

"The overall behavior of schizophrenic patients is primarily characterized by abnormally distorted perceptions of *what is real and what is not.* Some patients hear voices. . . . They believe that their ideas of the world are imposed on their minds by *outside* forces, and thus overcome, they are unable to separate fact from fancy. . . . At the same time, schizophrenic patients are *unable to generalize.* . . . This inability to make generalized abstractions is said to arise from an extremely *'concrete'* way of looking at the world. . . . Still others show periods of *extremely disruptive aggressive behavior* and require restraint to avoid *hurting themselves and others.* It seems clear that a disturbance of the thinking process occurs in all of these patients. But it is not at all clear that the same, unknown cause is the source of the problem in all cases, or that these extremely varied clinical problems can have the same biological basis. . . . Psychiatrists long found it difficult to understand the relationship between the 'positive symptoms' that are suffered by some schizophrenics – hallucination, thought disorders, and *delusions* – and the 'negative symptoms' expressed by others – loss of emotional responses, *inanimate postures, loss of spontaneous speech* and general lack of motivation." (Pages 260-261)

The second selection, which is from *Behavioral Neurology*, by Pincus and Tucker, and in which the emphases are again mine, is as follows:

"1. Auditory Hallucinations
   a. Audible thoughts (*voices speaking patient's* thoughts aloud)
   b. Voices arguing (two or more voices arguing *usually about patient – refer to patient in third person*)
   c. Voices commenting on *patient's* actions
2. Delusional Experiences
   a. Bodily sensations imposed on patient by some *external* source
   b. Thought being taken from his mind
   c. Thoughts ascribed to *others*
   d. Diffusion of thoughts (patient's thoughts experienced as *all around him*)
   e. Feelings, impulses, volitional acts imposed on him or under the control of *external* sources . . . . " (Page 62)

Pincus and Tucker discuss the genetic question at considerable length, and, after citing a number of studies, conclude:

"These studies do more than merely offer support for a theory about genetic influence in schizophrenia. They indicate that *it is a genetic disease.* They offer no support for the view that psycho-social environment plays any role in determining the risk of developing schizophrenia in individuals who are generally at high risk. The child of a schizophrenic has the same chance of developing the disease whether he is raised by his schizophrenic parent or in a normal environment." (Page 89)

In view of the quite substantial correspondence between the symptoms that one can predict by using the assumption of a narration complex and failures in it, and the symptoms actually observed in schizophrenia, it seems reasonable to conclude that the narration complex must exist as a separate, identifiable segment of the human brain, and that schizophrenia is brought on by departures from normal biological function that are induced in this complex by genetic mutations that are transmitted from one generation to the next.

This leads to two other aspects of schizophrenia that are conveniently introduced by the following additional quotes from *Behavioral Neurology* in which the emphases are again mine:

"Schizophrenia is primarily a disease of *young people.* Kraepellin noted that most patients were *under the age of thirty-five years* at the time of diagnosis, a finding which has been confirmed in many subsequent studies. The first clear-cut symptoms appear *before the age of twenty-five* in 50 percent of the cases; onset after the age of forty is unusual (Kraepelin, 1925). Symptoms rarely begin in the first decade, but when they do, they virtually always occur in the latter half, never before the age of five. *Childhood schizophrenia* has often been *confused* with *infantile autism*, a condition which usually begins in the first year of life and *always appears before age five. . . .*" (Page 60)

"*Disorders of speech* are the hallmark of early childhood autism, a behavioral syndrome first described by Kanner and *often mislabeled* 'childhood schizophrenia.' . . . It begins in the first few years of life. . . . Autistic children have a marked inability to form human relationships and give a sharp impression of extreme solitariness. . . . The most convincing evidence that the two conditions [autism and schizophrenia] are not identical is *genetic. There is no increase in the prevalence of schizophrenia in the parents or siblings of autistic children.* . . . Autistic children often have varying deficits in comprehension, symbolic thinking, and the formation of abstract concepts." (Pages 124-5-6)

Going to the beginning of the first of the two quotes, it is interesting to notice the difference that an evolutionary treatment of the origin of schizophrenia can bring to the view that it "is primarily a disease of young people [because] the first clear-cut symptoms appear before the age of twenty-five in 50 percent of the cases; [and] onset after the age of forty is unusual." For, if we imagine the situation of the first humans in which a narration complex had emerged, and who were then able to suffer from schizophrenia, the average life-span was almost certainly no more than about twenty years, and so schizophrenia, which would show its symptoms at age twenty-five, would have been very much a disease of *old age*. Thus, it could be that, in this disease of some of the "young people" of today, we can see a dim genetic marker of the approaching end of what must have been, at best, the relatively short life of our nearest evolutionary ancestor.

We can come on another of these age-related aspects of schizophrenia by noticing that if, as I believe should now be clear, the illness is a manifestation of inadequacies of one sort or another in the functioning of the narration complex, then we might expect to find what I would call "normal, developmental schizophrenia" in *young children*, during the period when they are capable of single-sentence speech *only*, and which I take to be an indication of the incomplete development of their narration complex. Evidently, such a normal manifestation of incomplete development of the brain of the young child would gradually disappear in the course of normal maturation.

I therefore believe that much of what are regarded as episodes of "childish behavior"—including delusions, the inability to cope with complex abstractions, episodes of cruelty, long series of "why" reflecting an inability to penetrate mystery—are simply manifestations of normal, developmental schizophrenia, which ultimately disappear in the normal child, as its narration complex becomes fully mature.

And it needs to be stressed that such developmental schizophrenia is hereditary only in the almost trivial sense that *all* normal biological development is hereditary, and so one would not expect to be able to find any particular evidence of schizophrenia among parents or siblings, linked to such normal development of almost every child.

However, the course of *normal* development of the narration complex might be *arrested* for a variety of purely *accidental* reasons, such as brain-damage suffered at birth, or simply a blow on the head, which have nothing to do with genetics and heredity. When normal development of the narration complex is arrested in some such non-hereditary way, the child so stricken will then be shackled to a condition of *permanent* schizophrenia, which displays almost all the symptoms of the schizophrenia of "old age," but which has absolutely no connection to parental schizophrenia. And I add "almost," since such schizophrenia in children will lack the great mass of narration-based experience on which the illness can draw in adults, and which will tend to give the symptoms in them a broader and even more perplexing quality.

I believe that this kind of *accidentally* induced *permanence*, in what is normally the passing, developmental schizophrenia of childhood, gives rise to the illness referred to in the two foregoing quotes as *"autism."* If you read them once more, you will see how nicely such an explanation fits their facts, as well as how difficult it would have been, in the absence of an evolutionary sketch of the development and functions of the narration complex, to avoid the confusion in interpretation which is now evident in them. "Childhood schizophrenia" is not at all a "mislabel" for autism; on the contrary, this label reflects the well-developed, collective intuition of the observant clinicians who tried to make it stick.

This has now become such a long digression on schizophrenia that it seems desirable to present a short summary of the implications of what the present chapter has had to say, so far. The picture we arrive at is one in which the creature that preceded us, as the storytelling playals that we are, must have spoken single sentences only, structured according to the patterns of an active-passive grammar. Lacking the capacity to narrate, with all that this can now be seen to imply, the behavior of the creature must have been somewhat like that of the human child, still displaying normal developmental schizophrenia.

The survival benefits of narration are so great that the behavior determined by the associated complex slowly came to dominate more and more of the entire behavior of the creature. The process must have been slow, because of the enormous increase in memory that is required to cope with the streams of second-hand "experience" associated with narration. As the creature continued to evolve, it continued to display developmental schizophrenia in its children, but the normal course of development came to be that in which the power of narration captured the behavior of the mature creature. Thus, the mature period of its life was spent in the exploitation of the benefits of narration and its associated powers.

It seems reasonable that the brain of this early "human" would not have been able to support the immense demands of narration for more and more memory, and so, as a normal feature of its old age, would slowly have regressed to a condition of increasingly schizophrenic behavior, as its narration complex began to fail, and to resemble more and more that of its early childhood. However, the combined effects of further slow mutation, and the power of natural selection, linked to narration itself, gradually extended the capacity of the aging "human" brain to produce increased memory. Thus, a line of humans developed who were able to survive longer, without suffering the previously normal regression to schizophrenia, leaving other afflictions to usher in the ends of their lengthening lives.

But traces of the earlier "human" condition rest in the genes of the general human population, so that, today, some fraction of it remains vulnerable to relapse into schizophrenia at a time that seems now to be *late youth*, but which was really *old age* at the time of the emergence of narration. Viewed in this way, we can understand the anguish inspired by autistic children, growing into adults, while accidentally transformed into a hazy glimpse of our pre-narration ancestors.

As you will recall, it was possible to show that the three previous evolutionary stages—pre-playal, primal-play and game-play—were associated with identifiable parts of the brains of playals. It would therefore be reassuring if some place could be identified in the human brain with which the narration complex could be identified.

EVIDENCE FOR EXISTENCE OF A NARRATION COMPLEX

What I shall do now is extend the previous quote from *The Brain*, by Restak, to show that there is a place in the human brain that can be associated with the narration complex. As usual, the emphases are mine in the following quote:

> "The third 'brain', the cortex is most highly developed in humans. It is a kind of problem-solving and memorizing device to aid the two older formations of the brain in the struggle for survival. Dr. MacLean compares the cortex to a 'computer' that can look into the future and anticipate the consequences of actions. The cerebral cortex furnishes us with our most human qualities: our language, our ability to reason, to deal with symbols, and to develop a culture. The *prefrontal* areas of the cortex are the most highly developed. Dr. MacLean considers the development of the *prefrontal fibres* the most auspicious turn of events in the history of biology. 'It is this new development that makes possible the insight required to plan for the *needs of others* as well as the self, and to use our knowledge to *alleviate suffering* everywhere. In creating for the first time a creature with a *concern for all living things*, nature accomplished a one-hundred-eighty-degree turnabout from what had previously been a reptile-eat-reptile and dog-eat-dog world.' . . ."
> (Restak, pages 136-7)

Evidently, the place in the brain that I am looking for is known as the "prefrontal areas of the cortex." This is clear from the references by Dr. MacLean to a number of modes of behavior which express *empathy*, and which therefore link to animation and hence narration. It is also significant that the prefrontal areas are identified as a "new development." Support for this conclusion, of a quite different kind, comes from recent comparisons of brain activity in normal and schizophrenic individuals. These comparisons were made using scanning devices capable of identifying the level of chemical activity in parts of the functioning brain. What is found is that schizophrenics show very clear deficiencies of chemical activity in the prefrontal cortex.

It is therefore possible, at this point, to identify not three, but four spaces in the brain that can be associated with four major, play-related evolutionary stages leading ultimately to humans. These are:

1. **Expansion of the upper brainstem;**	Pre-playal instinctive behavior, as in reptiles; no play;
2. **Limbic system;**	Primal play, as in mammals; learned-routine space; copy-learning; felt selfness; understanding; abstraction; REM sleep;
3. **Cortex;**	Game-play; active-passive, single-sentence language; self-challenge; unconscious life; madness;
4. **Prefrontal cortex;**	Narration; animation; peculiarly human behavior; schizophrenia, autism.

The comments in the right-hand column are just brief reminders of the behavioral situation corresponding to each of the four new evolutionary developments of the brain, as outlined in this and previous chapters. And it is important that these be taken to be four "spaces" rather than four "places," since they must be interconnected, and therefore spread over and into different places in the brain, merging as they must.

We can see from this that the new developments leading to the fourth major part of the brain must have led to a new species, since the changes that evidently took place in the prefrontal cortex are too extensive to have allowed successful mating between those who had come before and those who came after the changes. And the new species must be us, since the changes account for the last significant development in our behavior.

It is also clear that, at the time of our final emergence as a single species, we were starting to tell stories, and not just shouting: IAY-YAI. Consequently, we can begin to see the possibility of the stories that were handed down from generation to generation having among them some that recount, maybe dimly, the story of this last upheaval which produced the animal species that is ours.

There must be, among these stories, some that tell about the disappearance of every living trace of the creatures just one step back, and which, at the start, were all around the first few of us; stories that tell about the ancient ones who could make a sentence and play games, and hum a melody, and learn from their parents. They even dreamed dreams, but, and here's the catch, couldn't share them; nor could they take suggestions.

Seems they paid a high price for just being themselves, humming little tunes, and minding their own business. But it's not taking our suggestions, I guess, that really did them in, for our ancestors must surely have suggested some useful ways for them to be our slaves, and so avoid extinction. Or maybe not; perhaps it was just the still inexperienced psychoanalysts of the time, who, having come on suggestion and hypnosis only very recently, simply drove the poor creatures clean out of their unyielding, single-sentence minds!

TIME OF HUMAN EMERGENCE

It would be desirable to attach all the foregoing evolutionary developments to some general chronology which relates to the progressive physical transformation of the earth. One important reason for wanting to do so is the fact that I intend to discuss some of the consequences of the general dispersion of the human species over the surface of the earth. And since the continental and other major surface features of the earth have shifted and changed over time, it is desirable to have some timescale associated with what has been talked about, if we want to be sure of what these features were like when the human dispersion took place.

The present evidence is such that I associate the human storyteller with the "cro-magnon" human which appears to have established itself in Europe some 60,000 years ago. But although the name "cro-magnon" is taken from the name of a place in France where fossils of this creature were first found, there is evidence that this final form of human originated elsewhere. Two questions arise naturally: where did this final form of human originate, and when?

Until just a few years ago it was considered possible that the cro-magnon human evolved from the "Neanderthal" pre-human. The Neanderthal appeared about 100,000 years ago, that is, some four thousand generations back. Compared to ourselves, it had a very massive bony skeleton, with individual major bones whose diameters were something like twice those of ours. The bones of its skull were equally massive compared to ours, with relatively large ridges surrounding the eyes. So, did the cro-magnon evolve from the Neanderthal, that is, have we descended from Neanderthals?

It seems almost certain that this could *not* have been so, and here I shall give only those reasons which come from the argument that I am advancing. In particular, I am asserting that the last major transition from pre-humans to humans was the development of the narration complex. This clearly called for major additions to the previously existing parts of the brain, and these would, I believe, have taken many more generations to evolve than the less than four thousand that would be available between Neanderthal "ancestors" and ourselves.

In addition, it should be noticed that the volume of the brain of the Neanderthal was slightly *greater* than that of ours, and it would seem unlikely that the *addition* of the enormous complexity of narration to the brain would be accompanied by even a slight *reduction* in its volume.

It therefore seems that rather than being descended directly from Neanderthals, we are more probably descended from some common ancestor. This common ancestor would have given rise, perhaps forty thousand generations ago—a million years back—to a number of streams, of which one led to Neanderthals and one to us. Such an origin would provide the increase, from four- to forty-thousand, in the number of generations that must have been required to go from the single-sentence, here-and-now language of our predecessors to the storytellers that we have become, with all the enormous complexity that, as I have indicated, this implies.

The new storytelling species probably emerged, in more or less the form in which we can see ourselves today, some hundred-thousand years ago, that is, four thousand generations back. We would then have emerged just a little later than Neanderthals, but more in parallel with them than in series, spreading out as they had, and, it would seem, ultimately driving them to extinction as we went.

However, even such an origin of the human species, taking us back a hundred-thousand years or so, would not go back to a time when the relative positions of continents and oceans were significantly different from what they are today. What changed during this time were the surfaces of continents and the depths of oceans, as great mile-high sheets of ice would cover the northern parts of Eurasia and America, last for tens of thousands of years, and then melt and run back into the oceans, restoring their levels and covering previously exposed land-bridges between continents. The most recent of these ice-ages came to an end about fifteen-thousand years ago, that is, about seven hundred generations back.

ORIGIN AND SPREAD OF THE HUMAN SPECIES

The foregoing allows us to have some idea of when the human species might have emerged. But where did we first emerge? Where did this physically rather frail, storytelling creature make its first appearance? This question has recently taken on new vigor as the power of modern molecular biology has become involved in the answer. As things now stand, two possible locations have been suggested: one in east-central Africa, and the other in central Asia. The details of these approaches need not concern us here, and I am going to assume the African location in what follows.

Proceeding from this assumption, the picture of the spread of our species is one in which, starting from an African origin, some four-thousand generations or so ago, what has come to be known as the cro-magnon human radiated out over the surface of the earth, replacing all previous varieties of near-human creatures. This implies that the spectrum of the "racial" types that exist currently within the human species consists of specializations of the original African human.

As we have seen, when this spreading started, the continental land masses had long ago taken up the positions they occupy today. But, as the spreading proceeded, great mountains of ice formed over the northern half of Eurasia and North America. These mile-high mountains of ice were to retreat finally only some fifteen or so thousand years ago. Thus, most of the conditions of early human existence were those determined by this most recent ice-age, and only about the last fifteen-thousand years, or the last seven hundred generations, have been lived in physical conditions approximately similar to those we experience today.

While the ice-sheets persisted, a considerable amount of what is now water in the oceans was trapped in them. Thus, the human groups moving out of Africa had extensive land-bridges across which they could move into Europe and Asia. Further afield, the human group moving east could cross what was then a land-bridge connecting eastern Asia and North America, and descend subsequently into South America. As well, a group moving east along the southern fringe of Asia could move out into what are now the extensive island-chains of the western Pacific. In this way, our species spread across the surface of the earth.

So what are we, that we could have spread to everywhere? We are unique in our deadliness when awake; probably unique in the clarity of our dreams, but this is unsure, since we are unique in describing dreams; unique in the awesomeness of our madness; unique in the expression of our inescapable mercy; unique in the depth of our fear; unique in our willingness to continue to follow a single stranger; unique in the execution of suicide. But, above all, we are unique in the tenacity with which we hold to self-deception and false pride, for, when the almost infinite biology of our brains makes selections out of tables based on past pain, written just in molecules, we say, with our words, that *we* have "chosen," by which we mean that "it" might, arising in ourselves alone, have been different. When it is *we, one by one,* who continue to *be* chosen, by our extending pasts, made up of a newly-useful life-bound past, which stands on a more distant biology stretching back and back, to what, in early times like these, we say are Gods, but what we shall, some day, come to say were simply rustlings in . . . nothing.

THE EVOLVING HUMAN

Where there is room for possible biological specialization, evolution will explore the room using living entities as the vehicles of exploration, and their survival or extinction as a measure of the amount of available room. So having arrived at the human species, evolution continued to explore the possibilities of specialization. What, then, are the possibilities?

In the first place, there is the entire biological system that preceded the emergence of the human species, and on which our basic existence is based. Thus, as humans moved about, exposing themselves to new physical environments, minor evolutionary changes could occur beneficially in those parts of this basic biological system that interacted most directly with the various physical environments. Evidently, such changes did occur since, although humans belong to the same species, and must have originated in a small homogeneous group, we presently embody numerous physical variations, as are apparent in the differences—in skin-color and hair-type, for instance—to be found among us. This sort of physical change, in response to changes in physical environment, was certainly a way in which evolution could, and, evidently, did do its specializing.

There is also a second way in which specialization can proceed, and that is by acting on the narration complex, as the last evolutionary transition leading to humans. Because this last transition is also embedded in biology, and, as surely as it was established by evolution, it can continue to be moulded and specialized by it.

But evolutionary changes here would differ quite markedly from those that affect mainly external, physical characteristics, because the narration complex is the source of every mode of *behavior* that we regard as being uniquely human. This includes everything from schizophrenia, to our ability to construct theories; from our ability to tell stories, to our willingness to accept others, especially different others, into our lives. So it would take a larger book than this to explore what even minor differences in the evolutionary specialization of the narration complex could have meant, for hereditary differences, in the human-specific behaviors of the branches of our species, which wandered off to separate corners of the earth.

Of course, I know how anxiously racists of every "race" await confirmation, and how happily some of them will seize on this last paragraph. Let me say, therefore, that I see no reason for pretending that what seems to have been a reasonable set of possible evolutionary specializations could not have occurred, or did not occur. However, I suggest to the more impatient racists that they wait for a fuller exploration of the possibilities—by evolution *herself*, for instance. She comes well recommended, not having ever committed an error. She knows where the fossils are buried, especially those of the *great masters* of their times and places, now known to her alone. So now she's too shrewd to place her bets . . . before the game is really over.

CHAPTER IX
HUMAN OCCUPATIONS

THE PROPHET

Perhaps, if we are going to examine the occupation of prophet, one should start with Isaiah. But there are other ways, and I shall start with a short movie of a frog catching a fly. It runs for three seconds, and consists of a sequence of ninety pictures, numbered in the order in which they're made. The movie shows the fly alighting on a leaf, and the frog, in a flash, extending its tongue and capturing it.

If, instead of running the sequence as a movie, we place ourselves at picture 41, for instance, we would say that picture 40 is located in the "past." But what of 42? We would say that, when 41 was made, 42 was "yet to happen"; as we say, it was in the "future" of frame 41. And when finally picture 42 materialises, we say 41 is in *its* past.

What we can notice is that the world of the frog, the evolutionary world, is situated at the edge of a continuously extending stream of the "already-happened," the "already-lived." Stored in the genes of the frog, as in ours, is such a long record of this "already-lived," that we and the frog are, inescapably, expressions of the seeming endlessness of the extending of the stream. Built into the frog, and into us, is the seeming inevitability of some "yet-to-happen."

Thus, for the frog, the past dissolves into an inexorable future; indeed, the frog is simply a biological accumulation of the evidence of such persistence. And so the apparently simple act of starting to thrust its tongue at the fly is much more than just the initiation of an instinctive act intended to satisfy hunger and guarantee survival; in addition, *it is an instinctive act of prophecy*, driven, so it would seem, by the evidence stored in genes, according to which flies have always persisted into the "near" future.

That the frog should instinctively prophesy would seem, from the evolutionary point of view, a curious outcome, since, as I have indicated on a number of previous occasions, evolution, as a process, is not directed at the future. Indeed, nothing would so weaken an argument, which claimed to be evolutionary, as the postulating of a feature or process that developed supposedly because it was aimed, for example, at the future survival of some line of creatures. What is curious, then, is that, in spite of this aversion to inclusion of "future" as a determinant of development, a relentless process of natural selection should lead to the embedment in genes of such a powerful expression of the persistence of the environmental states in which evolution proceeds, and that even elementary instinctive behavior should come to express such an onward continuity of, and with the present, as *almost* to invoke "the future."

Of course, what we see in the frog is far removed from what we can hear of Isaiah, or, for that matter, hear directly from economists, tea-leaf readers, meteorologists and other human prophets. And so the question arises as to how to cover the evolutionary distance stretching from frogs to economists.

Frustratingly small as this distance might seem on occasion, I shall cover it in steps corresponding to the major evolutionary stages that, as I have shown, followed the reptilian stage, and which are still clearly represented in the structure and functioning of the human brain. Where I shall be leading us, then, is to the conclusion that the human playal is sometimes prophesying instinctively like the frog, using its upper brain stem, sometimes like a post-reptile playal using its limbic system, sometimes in ways that reflect the functioning of its cortex, and sometimes in ways that reflect the uniqueness and power of human narration itself, arising in the prefrontal cortex.

We can begin by recalling that the post-reptile playal can, employing its limbic system, execute behavior based on copy-learning, which reflects extended passages drawn from its own life-experiences, that is, from its individual past, including its recent past. Now, if we think of these recent episodes as stored physically side by side (in space allotted by REM sleep), as they must somehow be in a limited part of a small brain, we can imagine that the space between the stored episodes would become filled with new "episodes" produced by the "diffusion" of portions of the real episodes into the intervening space to produce new, mixed "episodes" that the playal would recognize as being similar to some actual past episode, but different just the same. (I am using "diffusion" in this model rather than the growth, spreading and interconnection of hundreds of millions of neural components, such as dendrites and synapses, since a model based on such microscopic components and activity involves much complexity but contributes little to imagining how new "episodes" might be formed.) These artificial past episodes give access to a multitude of artificial onward "continuations" of the segments and residual traces of actual recent pasts that reside in them. Of course, these are continuations of the recent pasts that "have never been," but among which at least some could come to be. Such artificial continuations of the recent past harbor possible futures.

So now the playal is no longer limited to instinctive acts of prophecy like those of the frog, which only express a constrained future derived from a single stream of actual, distant pasts, riveted to the immediacy of survival, as filtered by natural selection and stored in its genes. Now, in addition, the playal can prophesy based on a profusion of artificial "pasts," synthesized in the beating blackness of its brain from passages in its own, actual past.

Compared with the frog, the playal prophet is now immensely free, since it can express, in its present behavior, features that only dimly couple to any actual past. In particular, this means that its ongoing behavior at any time might reflect a temporary detachment from the urgencies and immediacies of survival, since, what might now be driving the behavior of the playal is an artificial extension of what it has actually lived, that is, a "future" that has *no actual past.* So now the playal is free to prophesy that the future might be greatly different from extensions of the actual past, unlike the frog, which, except for minute genetic mutations which might occur from one generation to the next, must prophesy, as its parents did, of a narrowly constrained future-from-the-past, indeed, of a future-like-the-past.

This freedom to create artificial continuations of the recent past, which are, in effect, latent prophecies, can be seen as forming the natural basis for the beginnings of both *tool making* and, later, in immediate pre-humans, the *burial of the dead.* For the brain that can compound artificial futures by mixing and extending actual pasts can construct a future in which some place, that didn't happen to have a convenient supply of well-shaped stones for hammering with, can come to have such stones—even if it involves carrying them from a place that has stones, to the place that doesn't, in a burdensome act of prophecy. The frog sets out to bring the fly to the place where it is needed, prophesying to itself that they will both persist; the pre-human creature sets out to have stones where they are needed, prophesying to itself that the place, the stones and, most of all, its own self will be there.

Here we can see how prophesying to itself can lead to the appearance of aimlessness in the behavior of a playal, if we picture some ancient ancestor plodding along, with stones carried awkwardly, as it enacts, for itself, some prophecy of a place without stones turned into a place with. This act of seemingly aimless carrying, simple as it might appear, is doubtless one of the most far-reaching instances of prophecy as action. For, clearly, the creature is responding to some prophecy which it has made to itself, and in which the immediate inconvenience of carrying stones is overcome by some prospect as realized, as of then, only in an artificial "future," running in a sequence in its brain. It is the growing gap between the making of the prophecy, and its realization outside the brain of the creature, that lengthens the periods of seemingly aimless activity. It is this same power of prophecy in action which, later on, will drive on through the almost inconceivably aimless "idleness" to be seen when the creature begins to *shape* the stones by striking one with another, oblivious to the immediacies of its environment.

In a similar way, we can see that placing the dead in a "safe" place, accompanied by implements and food, is a production of the same brain, prophesying life after death. Given the extensive individual experience which pre-humans would have had of birth, living, the need for food, as well as sleep and dying, not only in humans, but in other creatures around them, it is not difficult to see that, among the *pastless* futures that could be compounded in the evolving pre-human brain, and be expressed in acts of prophecy, would be living-after-dying. Although unable to express prophecy in narration, as with us, the pre-human was no less well equipped than the frog to express prophecy in forms of behavior other than storytelling.

It would seem unlikely that the practice would have begun *without* the inclusion of food, for instance, with the body, since, although the prophecy of life after death is not easily falsified, certainly life without food *is*. So it is not surprising that pre-humans did, at some point, begin to bury their dead, and it would have been surprising if they had done so without providing food and other "necessities," as everyday existence would have suggested, and of which even the most fantastic prophecy would take account.

Evidently, our "advanced" north-American version of this prophecy has shifted, since it would be regarded as downright bizarre to place a handful of raisins in my friend's coffin. Indeed, our version has shifted to the social left, in that there is now assurance of sustenance where we're supposed to be going; what is needed is proper dress to get in!

How could a creature, which lives by hunting and gathering food, ever survive such lengthening periods of "idleness" and inattention to its immediate environment? The answer, of course, is . . . uncertainly. For that is the true nature of how survival is affected by prophecy in action. Some paths mapped out by prophecy lead to survival and further prophecy, and some lead to destruction and extinction.

Thus, prophecies in pre-humans resemble mutations in genes; they map out new programs of behavior which find expression in the future. But genes are part of the "bio-ware," hard to change, even when things go wrong, while prophecies compounded of actual life-experience are part of the "software," with countless behavioral options that are relatively easy to change when things go wrong—perhaps too easy, as I shall explain shortly, when they are not constrained by a particular condition that emerges only later in the fully human prophet.

So let us go now to what narration might add to the role of prophet. What might be the effects of the power of narration on prophecy? Some, at least, of these are not difficult to imagine, since narration adds to the exclusively individual past of the pre-human playal whole families of the narrated pasts of others which, when animated and rendered expressible in the behavior of an individual human, provide an extensive new substrate out of which to synthesize more artificial futures for which there are no actual pasts. But this is really only a *quan*titative shift in the human compared to the pre-human.

More interesting is the possibility of the *quali*tative shift that comes about as I shall now explain. Imagine a human "A" that synthesizes an artificial future (a'f) and conveys it by *narration* of a prophecy to human "B." Human B then uses (a'f) to synthesize a new artificial future and prophecy which could be written (b'a'f), since B can process A's prophecy *as if* it were an actual past for B. If we add humans C and D we can arrive at an artificial future and prophecy (d'c'b'a'f), and so on. This ability of a group of humans to employ narration to produce artificial futures by cascading the ability of each individual to prophesy is *quali*tatively new. But it is not just new, it is also immensely powerful

To see the source of this power, we need to be reminded of the value for survival of prophecies that turn out to match the single, real, future as it unfolds. Because of their vital importance, I shall refer to them as "veridical" prophecies, where the adjective "veridical" is used to suggest that these matching prophecies "tell the truth." (Later in this Chapter I'll show why a prophecy can never be better than "almost" veridical. However, for now, I'll continue as if a prophecy could actually be "fully" veridical.)

It could be that (even just "almost") veridical prophecy is at the root of human survival, and especially those aspects of our survival that reflect an increasing capacity to redirect natural forces in the service of human prophecy. If, for example, one obtains food by hunting and gathering, then veridical prophecy improves the accuracy and hence the ease with which we can find the animals we chase, the plants we gather, and the water we drink. If we are farmers, veridical prophecy tells us which plants will perform better, and where. If we are NASA, veridical prophecy tells us how and where to fire a rocket so that it will pass, a year or so later, at a given distance from a rapidly moving sister planet, millions of miles away. The next time you listen to a weather forecast, and have your raincoat handy to protect you from the rain, thank veridical prophecy; the other times, well . . . better to read on.

So let us return to the cascaded prophecy (d'c'b'a'f) to see how it comes to relate to veridical prophecy. The relationship is easy to appreciate, because it relies on the fact that the prophecy can be modified by the cascaded efforts of *many* prophets, rather than by the efforts of just one, and so brought relatively "quickly" into a veridical state. Further, this cascading can continue *from one generation to the next*. It is for this reason that the human prophet, fully endowed with narration, became qualitatively different from even the immediate pre-human prophet, and, without a little thought, might be supposed, mistakenly, to have no links at all with frogs as prophets, so vast has the gap become between us and them.

But there is a link, and a sobering link it is too. To see what the link might be, we can begin by noticing that, because of the long and unforgiving filtering provided by natural selection, the instinctive prophecy of the frog is almost always veridical. Its very continuing survival is good evidence of this. By contrast, most human prophecy is, in the first instance, at the level of the individual, non-veridical. This is so for two reasons. First, to the extent that instinctive function has been replaced with learned function, all the veridical prophecy that would have been associated with vanished instinctive functions has vanished with them.

Second, and much more significantly, since much of our voluminous prophecy is derived from futures that can be only weakly coupled to actual pasts, most human prophecy will be the expression of baseless fantasy, for which there could simply be no eventual connection to any actual, emerging future, since an actual future cannot, like the futures in our brains, be almost arbitrarily disconnected from actual pasts. What links us back to the frog, as constrained, veridical prophets, is the possibility of the filtering and tuning of our prophecies in the numerous cascaded and parallel chains of human brains and bodies, loosely coupled by *narration*. In this way, until now, at least, the waves of human behavior running out of individual, non-veridical prophecy have been more than cancelled by the waves running out of communal, veridical prophecy.

It is interesting to notice that the human veridical counter-wave, driven as it is by the power of narration, would not have been available even to quite immediate pre-humans who, as a result, would have been exposed to modes of individual behavior driven by the wildest of prophecy, unchecked by a relatively fast-acting process of cumulative, communal testing and modification. Such an evolutionary experiment would seem to have lacked an important survival feature, in that it would have largely outgrown the instinctive capacity for veridical prophecy of the frog, without having yet donned the kind of substitute protection, albeit rather frilly and fragile, with which narration seems to have provided us.

We can see here the origin of what we know as "belief." For a prophecy that has been shown to be veridical over and over again, because of its importance for survival, develops a certain special relationship to the creatures that entertain such a prophecy, and comes to determine their behavior. This special relationship we call "belief." Thus, the relationship of the frog to its prophecies is simply instinctive belief, as unavoidable as the prophecies are veridical. But in the human case there are three alternatives:

1. Prophesies that have been shown to be veridical;

2. Prophesies that have been shown to be not veridical;

3. Prophesies that have not been shown to be either veridical or not veridical.

Prophecies as in (1) we share with the frog, and we relate to them with "belief." Prophecies as in (2) we share mainly with the extinct antecedents of the frog, and we relate to them with "disbelief." Prophecies as in (3) are particularly interesting, in that survival requires neither belief nor disbelief, and excludes neither. When the relationship is one of belief, it is said to be one of "faith." Evidently, genetic transformations brought on by mutations imply enforced expressions of faith, since the departures in behavior, which flow from the mutation, must proceed in the context of new prophecies that the mutation will engender, even though, by the nature of the newness of the circumstances, such prophecies necessarily form part of alternative (3).

Belief, disbelief and faith are relationships inside the narrating brain which determine how a particular prophecy will affect behavior, and so will affect survival. It is therefore important to remain aware that the continuing veridical nature of a prophecy is related entirely to the extent to which those environmental conditions that made it veridical, in the first place, continue to prevail.

It is not difficult to see from all this that the human prophet must be a continually shifting and complex expression of instinctive prophecy, with an overlay of pre-human prophecy to oneself, and with a further overlay on that of truly human prophecy, filtered and uttered through the power of narration.

So, when some modern prophet climbs the grand stairway of the Lincoln Memorial, in the warm Washington sunshine, he brushes at a fly, as he goes, in an instinctive act of prophecy. He climbs up, all the while prophesying to himself about the effect he will have on the multitude assembled to hear the word. Then, having settled down before the microphones, he utters human prophecy, saying: "I have a dream . . . ," and goes on to paint a picture of a future America linked to its past, but different just the same. The picture he prophesies is derived from an array of possible futures, compounded from episodes in his own past, which have themselves been mixed with the prophecies of other prophets, like Isaiah, from the past.

Veridical?

I say: "probably"; and, so saying, have now involved myself with the truly strange notion of "probability." What gives rise to this strange notion?

Evidently, the notion of probability is closely linked to the fact that we can construct vast arrays of futures, that is, of artificial pasts, of which only one can form a veridical prophecy. But even though, in any given situation, only one prophecy can be veridical, we can often, as suggested by alternative (3) above, find a set of futures that cannot be excluded as being the basis of a veridical prophecy. Such futures we refer to as "possible" futures, by which we mean that there is nothing in our stored set of actual pasts, including the narrated pasts of others, that would exclude the emergence of such a new actual past, that is, of such a future. Thus, what we refer to as the "possible" is the set of futures out of which we are led, by the past, to entertain the "belief" that veridical prophecies could be drawn. So how does this help with getting at the "probable"?

It helps, because we can now do the following little mental experiment. Let us suppose that we have a "possible" set that consists of only two futures out of which two different prophesies pA and pB can be made. Evidently, both pA and pB can't be veridical, even though they are both possible. But since veridical prophecy is such an important factor in the survival of humans, it becomes important to try to know which of the two prophecies is veridical. And now we come on the core of the question of probability, because there is no evidence to suggest that humans can make a veridical prophecy about which of two possible prophecies will be veridical, that is, *we cannot prophesy which prophecy will correspond to the actual future.* If we view the two prophecies as narrations, then we are unable to tell which one of them will eventually come to contain an untruth, where "untruth" is as explained in the previous chapter.

It is because of this inability to prophesy which of two possible prophecies will correspond to the future that the narrating brain appeals to *probability*, and it is important to understand that this appeal to probability never overcomes the fundamental inability. But, since the appeal to probability is so widespread among humans, it must provide some benefit, and so it is important to see what this benefit might be.

A good place to start is back with our frog, whose prophetic power is based on the accumulation of behavioral routines by the process of mutations coupled with survival as dictated by natural selection. Thus our frog's power as a prophet is clearly based on the accumulation of its extended past (experience) as recorded in its genes. Furthermore, the individual frog can only continue to build up new pasts so long as these new pasts resemble the old pasts, the effects of which are stored in its genes. Thus, to the extent that the frog is a veridical prophet, it is a severely limited prophet, because it is constrained to prophesy that the next past, that is the future, will be the same as previous pasts. Consequently, it escapes extinction not because it has succeeded in "prophesying the future," but because it incorporates the past in a particular way, which has allowed it to cope with changes in its environment, as these have occurred . . . *so far.*

It is against this background that probability has to be seen, to appreciate its power on the one hand, and its fundamental limitations on the other. What probability does is give the narrating brain a new way of organizing the stored episodes originating in its relatively recent past, a way which is not available to the frog, and so it provides some additional capacity for responding to changes in the environment when these occur. Viewed in this way, probability can be seen as simply one of the properties of the narrating brain which serves to harness more effectively, for survival, the contents of its vastly increased memory, and its ability to compound artificial pasts from real pasts.

But what probability doesn't do is make even a dent in the inability to prophesy which of two possible prophecies will be veridical. In fact, the appeal to probability and probabilities is always marked by the abandonment of any hope of doing so, whether this abandonment is made explicit or not. Thus, when, previously, I replied: "probably," to the question as to whether a prophecy was veridical, I might have replied, instead, that I really didn't know what the answer should be, and, furthermore, couldn't know, but that my past leads me to believe, and so on.

What all this allows us to see is that probability is a pure creature of the human brain, endowed, as it is, with the capacity to narrate, and to prophesy in a form that includes narration. The frog, of course, has none of this. All it is constrained to do is express the long chain of real pasts, and the survival of its ancestors in this real past, as stored in its genes. Thus, at the level of the frog, there is no probability, since there is no array of artificial pasts to give rise to it. Indeed, the frog *needs* no probability, because it is confined to the kind of prophecy that flows out of a single, actual past, and so it prophesies that the future will be like the past.

This suggests, and very strongly, that a world without narrating creatures would be a world without probabilities. Indeed, what it suggests is that "explanations" of the world around us, and especially of the inanimate world, that present probabilities *as if* they were *intrinsic* parts of that world, are at best, *reflections of the origin and nature of narration itself,* rather than of a world apart from narration—as the world once surely was, before evolution stumbled first on play, and then narration, with its new and enigmatic prophets.

THE TOOLMAKER

Much emphasis has been placed, in the past, on the significance of the tool for the evolutionary emergence of humans. This is doubtless due to the fact that the only older human fossil-remains that are available to us are bones and teeth on the one hand, and stone tools on the other. This has tended to force the assembly of the human evolutionary story out of data reflecting this one aspect of the existence of our ancestors. The present discussion, proceeding from a basis in primal play, allows, as I have tried to show, the reconstruction of our evolutionary origins by tracing the development of a number of our features, including the development of the brain, and the progressive emergence of our particular type of human behavior. This has allowed me to escape what I consider to have been the excessive emphasis on "man the toolmaker" and, instead, to now position the tool in a larger context, extending from its likely first appearance to its role in our world.

Although we have a reasonably clear picture of what a tool is today, it is not quite so clear at which point, in the course of evolution, entities became tools. In order to establish a framework in which the discussion can proceed, I shall adopt the following definitions:

> A "tool" (or "instrument") is an entity that (a), is controlled by a creature in executing one or more of its behavioral routines, and (b), is not a part of the creature's biological endowment.

> During the time that an entity is related to a creature as in (a) and (b), the creature is said to "use" the entity as a tool.

> When an entity is transformed by a creature so that all or part of it can be used as a tool, the creature is said to "make" the tool from the entity that is transformed.

It is a clear implication of these definitions that the use of tools forms part of the behavior of even the most elementary living entity, since the controlled ingestion or absorption of any kind of foreign substance constitutes its use by such a living entity as a tool. Indeed, the use by living entities of components of their environments, as tools, constitutes the bedrock of self-serving evolutionary processes. So what is novel, from the evolutionary point of view, is not the *use* of tools but their *making*, and the conditions that would give rise to this.

To see what these conditions must be, we should begin by observing human creatures making tools. We can derive from this the following group of aspects of tool-making activity:

A. Although tool making is confined to the creature's waking state, it does not relate to the external situation defined either by its immediate past or its present;

B. It is executed by individuals, and sometimes in solitude;

C. It is not bound to any particular time, frequency, or place;

D. The tool—the outcome of tool making—has a significant "lifetime" not bound to that of the toolmaker;

E. The entities that are transformed to make tools can have a wide variety of starting locations, forms and compositions.

We can see from these aspects of tool making that, as mentioned briefly in the discussion of the prophet, the conditions for tool making must be those that arise from the ability of a creature to add to its set of instinctive, species-specific interactions with its environment, some other set based on its ability to compound other pasts than those stored exclusively in its genes.

The first opportunity that evolution presents for this is the acquisition of copy-learning which follows primal play, and so we can expect, as is the case, that tool making would not be found in creatures other than playals. Furthermore, we can expect that it would have emerged in the usual minute steps that characterize such emergence, beginning with some activity that is hardly distinguishable from instinctive activity, which it augments. And again, as mentioned in connection with the prophet, the initial "making" of tools must have consisted of activities no more spectacular than the carrying of stones from one place to another, that is, of almost imperceptible variations on their instinctive using.

From all of this it is easy to see that we were certainly not the first toolmakers, and that the tendency to present humanity as characterized by "man-the-toolmaker" is without basis. Of course, knowledge of the behavior of the chimpanzee, which includes the stripping of branches to make probes for invading the nests of ants, is quite general, but the tendency to look for and "find" great gaps between ourselves and other primates serves to limit appreciation of the elegant act of incipient tool making that this clearly is, even though the foregoing features (A) to (E) are not all fully met.

We can now combine this brief discussion of tool making, with what has gone before on language, to produce a rough picture of the emerging playal as this relates to them both. It seems reasonable that the facial signaling of game-play must have accompanied "carrying," and preceded the actual making of tools. When actual tool making emerges, it was followed by the beginnings of sound-based, single-sentence speech. This was followed by narration with its multi-sentence speech. This was followed by narration with its multi-sentence story, accompanied by the multi-part tool, which became the tools of the kind which we employ today.

During the two million or so years that it must have taken since the first efforts at actual tool making in materials as hard as flint, the "hand" that was to become the human hand must have undergone an extensive evolutionary development, as the self-serving advantages of tool making and tool-using took on evolutionary significance as it was exposed to the process of "natural" selection.

The mention of the human hand can, as you might know, introduce an interesting set of questions having to do with what is generally referred to as "handedness." These questions relate mainly to how a little more than nine out of ten of us come to consistently use our right rather than our left hands for doing all sorts of things, and what this means for the organization of the brain. Since such questions bear on much of what has been discussed here and in previous chapters, I shall look briefly at a possible origin of handedness.

If we start with the instinctive, external behavior of playals, there is simply no reason to believe that any advantage could come from favoring the development of one side of the animal rather than the other, that is, from external left-right asymmetry. On the contrary, one can imagine advantages arising from left-right symmetry in the animal, since predators are just as likely to come from the left as from the right. The same is true in the context of prey, since the opportunity for the food-securing lethal action is just as likely to present itself to the left as to the right side of the animal. Thus, at the instinctive level, there is no reason for evolution to lead to an externally, left-right un-symmetrical playal or, more especially, a species of playal which is un-symmetrical

I refer to an un-symmetrical species as particularly insignificant for evolution at the instinctive level since, even if, in the case of each individual animal, small deviations led to behavior that makes higher demands on one side than on the other, this would be due to small left-right perturbations in the basic symmetry. But, as a whole, the species, at the instinctive level, will display overall symmetry, and any tendency, in individuals, to extreme asymmetry would be limited by a process of continuous extinction. In effect, the species would manifest a balanced ambidexterity as its average behavior, with a limited spread of right- and left-handedness at the level of the individuals of which it is composed.

However, the beginning emergence of narration and the associated capacity for prophecy, which would support intensive *making* of tools, especially in materials as hard as flint, change this tendency to symmetry in a fundamental way since, even if one imagines a playal using stones as tools instinctively in a symmetrical way, there is simply no way in which a playal can *make* a tool by a symmetrical operation with both hands. If the playal is going to produce a particular shape in one piece of stone by hammering on it with another, the activity is definitely going to require consistently *un-*symmetrical behavior—one hand will have to do one thing, and the other hand something *else*, at the *same* time. Even the most primitive making of tools relies on a difference in the roles of the two hands, and so better tool making, and the related better chances of survival, can come from specialization which improves *simultaneous* effectiveness in each of the different roles of the two hands in tool making—holding and striking, or, passive and active.

Finally, one can notice that specialization and development of some function such as dexterity in a hand consumes space and nervous resources in the limited volume of the brain. Thus, a clear economy in the use of space and resources in the brain results from specializing and developing only one of the hands in just one of the two different roles associated with tool making. In this way, the asymmetry of handedness allows for the evolutionary development of the *making* of tools, with all its attendant benefits, at the minimum cost in brain-space and biological resources generally.

This is all that evolution needs to embark on a long series of narrowing specializations in which the playal would go from being fully symmetrical to being as asymmetrical as the advantages in the associated tool making behavior can sustain. In this way, narration and the associated tool making would lead, in a completely imaginable way, to handedness, as an evolutionary expression of the survival advantages of the made tool. I shall simply mention the possibility that, at about the same time in the evolution of the playal, the emergence of narration was involved in undoing the basic active-passive symmetry of the brain as this relates to language, and so was leading to another kind of "handedness" in the brain. This can hardly be just coincidence, but I shall explore it no further.

So now, why right- rather than left-handedness? Evidently, if we start with a symmetrical playal, then the genetic mutations that would lead to beginning specialization must lead to just one of the two possible forms of asymmetry, and the first kind to appear, be it what could become left- or right-handedness, will become the kind that is expressed in the specialization. Thus, the fact of being a right- rather than a left-handed species is due simply to some chance genetic excursion which was, evidently, an excursion to the right at the point in our evolution at which such a deviation could first help to amplify the advantages of tool making.

It is useful to notice that what we refer to as "handedness" is not confined to our hands alone, nor could it reasonably be, since the overall, external, left-right symmetry of the playal, clearly evident in all its external parts, is one of its major survival features, so that it must be controlled by an extensive genetic complex with widespread ramifications, and not just some isolated and minor genetic component. Thus, the kind of pronounced asymmetry that we find in the general displacement of humans toward (right-) handedness must be the result of shifts in such an extensive genetic complex.

Thus, I see no particular difficulty in accounting for the less than ten percent of us who are left-handed. In what must be an extensive genetic context, it seems reasonable to suppose that average right-centeredness in a species should include particular individual members who extend even further to the right than the average, as well as further to the left, yielding a spectrum of handedness going all the way from extreme uni-dextrous right-handers, to normal right-handers at the average, to the ambidextrous, to left-handers. The range from extreme right-handers to the just ambidextrous contains the roughly ninety percent of nominal right-handers. Although left-handers are easily accounted for in this way, one wonders to what extent some of their functions, other than those related directly to handedness, might have been weakened or enhanced coincidentally by a selection process favoring the right.

I should add that this genetically-determined off-centeredness will be present almost certainly in humans only, since we are the only (remaining) narrating toolmakers, and it was, in my view, only among emerging narrators that the benefits of tool making became sufficiently compelling to interact so strongly with the biological development of the playal as to disturb its basic external symmetry.

A number of quite different views of handedness and its possible origins can be had from the chapter entitled "The Puzzle of the Left-Hander" in the book *Left Brain, Right Brain*, by Springer and Deutch.

THE WRITER

One of the advantages of sound-signalling compared to grimace-signalling is the possibility with sound of being able to communicate even when we can't see one another. But there is a limit to how far apart we can be and still communicate even with sounds, partly because our vocal chords can generate only so much power, but also because there are times when we don't want everyone to hear what we're talking about.

Another aspect of sound-signalling—the fact that sounds spread out and quickly vanish—starts as being absolutely essential, and ends up as an inconvenience. It is essential since, if sounds didn't behave in this way, we couldn't use sound-signalling at all, for words would just keep piling up, one on the other; but it also ends up as an inconvenience if we want to *remember* what was said a few days ago. So it gets to be a great convenience if we can mark down on something what we said or would have said, and keep the something, or send it as quietly or as far as we wish.

I referred to "marking down" because that is what writing has come to be, but one can imagine that this sort of quiet signaling at a distance could, in the first instance, take a form quite similar to the carried tool. In effect, a particular kind of stone or shell could be agreed on as "standing for" some event, and this kind of object could be kept or sent, to record the passing of the event. Just such signaling was indeed used, and there were numerous variations on this theme. However, if one has tools, then *marks* on objects get to be a simpler and more flexible way of storing what was said, or would have been said, than the objects themselves.

What kinds of marks would one start with? Pictures, of course, fetched from the flow in the outer stream. Certainly we would draw extremely simple pictures at first, consisting of a stroke or two, but pictures just the same. The really crucial step, though, is the one in which a picture is made to stand for a word, rather than a whole message. From all the evidence, this invention occurred about five thousand years ago, in what is now Iraq. Moving toward the east, the identification between pictures and words remained, as we find it in Chinese and older Japanese. Moving toward the west, the pictures gave way to elements representing the *sounds of the syllables* in words. Finally, the syllabic representations were reduced to *alphabets*, of which the kind used for writing the sounds of English are an example. (Alphabets can also be found in soups and cereals, and made up into words, but these are reserved mainly for use by children and those few unfortunate writers who, having misjudged some part of the workings of evolution, must, in due course, eat their words!)

We should remember that speech isn't the only form of sound-signalling that gets written down, and so I shouldn't write about writers and forget composers. Of course, not all composers write, not any more than do all speakers write. We can hum a new melody quite happily without ever writing it down, in much the same way that we can speak a new sentence without ever writing it down. But some composers, fortunately, do write things down, and when they do, the writing they use is a special mixture of "syllables" and pictures. Some lucky people can look at written music and sing, the way some look at written speech and speak.

So far as the time-scale of all of this is concerned, we should notice that, compared to tool making, writing, or marking in general, is a very recent occurrence in the evolutionary stream. The earliest *made* tools occurred some two million or so years ago, while the earliest form of "writing" is no older than the final form of human narration itself, that is, about a hundred thousand years or so. What we would regard as "real" writing, still in a pictorial sense, goes back no further than about twenty thousand years.

THE INVENTOR

As I began by stating in Chapter I, a creature that continues to replicate, and so avoid extinction, must continue to meet environmental challenges. If we now follow the evolutionary process along the particular stream that leads to ourselves, we can see that there have emerged three ways in which living creatures can meet these challenges. The first way employs instinctive behavior based on the kinds of processes we find as late as in reptiles; the second makes use of behavior based on copy-learning as this appears in playals; the third employs behavior based on narration, as we find this in ourselves.

So what does all this have to do with invention? Quite a lot, and we can see this as follows. Evidently, the first of the three ways, involving instinctive behavior, does not allow individual creatures to cope with environmental challenges as their own life-experiences would dictate. However, with copy-learned behavior, creatures can begin to express their own life-experiences in behavior that copes with environmental assaults. In the most general sense, it is this self-serving expression of individual experience, as behavior for coping with environmental assaults, that constitutes "invention," and it is the emergence of copy-learning that confers on the playal the power of the "inventor." Thus, in all those creatures that are endowed with copy-learning, invention is already present as a form of behavior.

What then, if anything, did the development of narration and the emergence of humans add to invention? The answer to this important question is conveniently given in two parts. The first relates to the power of narration to allow individuals and groups that are separated by time and distance to continuously and communally modify existing inventions, and so produce improving ways of coping with environmental assaults. Thus, the effect of this aspect of narration on invention resembles that on prophecy, as realized in the cascaded prophecy (d'c'b'a'f), the difference being that, instead of (a'f) being prophet A's starting prophecy, which is modified by prophets B to D, it is inventor A's starting invention, which is modified by inventors B to D.

But this cascading can be viewed as just an extension of the general power of copy-learning and its sometimes beneficial imperfections. What is really new, and constitutes the second part of the answer, is the expansion that narration, but more particularly, its essential concomitant *animation* brings to the range of entities whose *"behavior"* can be copied, and so understood. *This* is the great qualitative change that creeps in with narration, for it is the capacity to embed in a brain a behaving phantom actor, and, at least as revolutionary, a behaving inanimate actor, which allows the narrating human to act out the behavior of the phantoms as *theories*, and to combine the inanimate actors with the stories of the phantoms to make theories of the behavior of the inanimate actors themselves. It is this that provides a whole new way of interacting artificially, in the human brain, with the "actors" of its environment, and of bringing them gradually within its control as new, "made" tools, that is, as *inventions of unprecedented power* in themselves, and hence as the basis of even more powerful cascaded inventions.

Thus, before the passage of what evolution would regard as a long time, the "behavior" of a phantom "gravity" can come to affect the "behavior" of an inanimate stone that falls toward the earth according to the "behavior" of an even more phantom theory, a story. Evidently, all these "behaviors" exist in the animating brain alone, but this is the only way in which it comes to understand, and to make its marvelous inventions, because, to understand, it must *copy behavior*, even if it must *create* behavior so as to copy it, just as the very first playals learned to copy just behavior, and so came to understand. In this way, within the narrating human brain, understanding and invention continue to support each other, as with any other playal, but it is *what* can be understood that is new, and it is this that makes the narrating brain, with its ability to animate, such a unique understander and peerless inventor.

THE ARTIST

We now have just about everything I need to relate art and the artist to the evolutionary world. To do this I am going to begin by placing tool making and writing in the context of what I shall call "past-condensing." To see what this means, let us begin with tool making, and notice that this can be viewed as a way of "condensing" the past into the inanimate world, that is, into the physical world. We can see this by noticing that, in making even the most primitive tool, the playal must express, in the tool, not only the ability to control itself as a generator of forces, that is, its deep, hereditary past, but must also express itself as a creature with a more recent hereditary and individual, experiential past. All of these pasts together are expressed through the combined instinctive, copy-learned and narration-based activity that is manifested in the tool.

Thus, tool making is an activity that, at many levels of behavior, entails a durable extension of the playal's past into the physical world, condensed out into what appears in the inert tool, as a unique physical object. It is this that gives the made tool its identifiable qualities as a made tool, and renders it distinguishable from those other items of the physical world among which it might be found, but which do not embody the past in this way which is unique to the evolutionary stream. Everything that has just been said of the tool can be said of writing, especially if we notice that writing includes drawing and the making of pictures.

We can therefore view tool making and writing together as approaching the outer limits of the evolutionary stream in which the condensed-out past has no possibility, by itself, of producing a next generation. Made tools and writings might therefore be looked on as having spread into those places in the evolutionary stream at which evolutionary replication, by these separate, extended condensations of the evolutionary past, ceases.

But even among tools and writings as non-reproducing "species," we have different varieties, since some, the "useful" ones, can be used to help transform the evolutionary world. These can also help to transform the physical world, and thus play an assisted role in the replication of other tools and writings. However, beyond the very extreme of usefulness, there are the "useless" tools and the "useless" markings which are not themselves used, and which represent the condensations of the evolutionary past into an ultimate, limiting, non-replicating and fundamental boundary, *beyond use.*

The activity that yields the objects that make up this fundamental boundary—beyond use—I call "art," and the objects that populate the boundary I call "art-objects." The creature involved in the making of an art-object I call an "artist."

In this way, we can see that the art-object must realize its character *as* an art-object not from *what* the object is, but from *how the object is situated with respect to the evolutionary world itself.*

I shall now summarize and expand the foregoing. The primary object of the expansion is to illustrate the quite wide sweep across the domain of art to which the foregoing leads, and to fill in the somewhat unusual context in which art is here placed by evolutionary explanation. The secondary object is to use the expansion to expose some novel extensions and connections of art. It is convenient to carry out both the summary and the expansion in the form of numbered paragraphs, as follows:

1. Art is the activity that yields the objects that populate one of the boundaries of the evolutionary world. These objects are known as art-objects.

2. The boundary in (1) is that at which the various levels of past in the evolutionary world condense out into the physical world and produce "useless," non-replicating objects, which are extensions of the evolutionary past.

3. Art-objects are "useless" in the sense that, unlike other "useful" tools and writings in the evolutionary world, they are not involved even in assisted, partial replication by transforming the physical world. They are *passive* condensations of the evolutionary past.

4. In view of (3), the activity in (1) is the limiting from of tool making and marking.

5. Art can be carried on by any tool-making creature, and so is clearly evident in ourselves, and was doubtless carried on by contemporaries of humans such as Neanderthals.

6. If Neanderthals indulged in art, they would have been limited to expressing their individual genetic past and their individual experiential past in their art-objects.

7. In our human case, we can add to the two past-streams in (6) a third stream derived from storytelling.

8. In view of (6) and (7), the nature of art-objects can be expected to have evolved as humans have evolved.

9. Art and art-objects will reflect condensations of the past related to the three streams in (7). They will therefore reflect features of the creature including "force-control" at one extreme, which we could associate with quality of execution, and, at the other extreme, unique personal and other life-experiences which we might associate with form, content, and the artist's "style." As art-objects go from

expressing mainly narration and its human complexities, at one extreme, to instinctive processes, at the other, they are said to become more and more "primitive."

10. In view of (9), art-objects produced by children will differ from those produced by adults with respect to execution, form and content. In particular, young children, since they have limited capacity for both copy-learning and storytelling, will tend to produce "primitive" art-objects.

11. In view of (9) and (10), adults who share common pasts, which extend into childhood, will tend to produce similar art-objects. Such sharing of pasts implies sharing common language, and so similar language groups will tend to produce similar art-objects.

12. Since an art-object reflects the condensed past of some artist, it is possible for the past reflected in an art-object to be partially shared with the past of an individual other than the artist. When the past reflected in the art-object corresponds to a (high-) no-pain segment of the past of the individual, the art-object is said by the individual to be ("ugly") "beautiful."

13. In view of (11) and (12), members of the same language group will tend to declare the same art-objects to be ugly, and the same art-objects to be beautiful.

14. Since, according to (12), beauty (ugliness) is related to shared pasts, it should be possible for one human to learn another's behavior related to the beauty (ugliness) of groups of art-objects by copy-learning and storytelling. This is analogous to learning a new language, and both types of learning are known to occur.

15. There is clearly a strong link between shared pasts, a shared language, shared art, shared art-objects and shared beauty. Individuals who shared all these would, evidently, have to share the same geographical region. A group of individuals who share all these aspects of their lives are said to be of, or to share, the same "culture."

16. Returning to (9), we can see that any particular art-object is uniquely related to the artist who made it, since it reflects the condensing of the unique past of the artist into the object, as the past of the artist stood when the object was made. It is, therefore, in a curious

but significant way, not possible to produce a "true" copy of an art-object.

17. Returning to (2) and (3), we can see that there can be no precise statement as to where the evolutionary boundary that is populated by art-objects begins, or where it ends, since it is not possible to say with assurance what is "useful" and what is "useless." Thus, art-objects merge imperceptibly with "useful" tools and markings on one side, and, on the other side, with the physical world, where it has only marginally interacted with the evolutionary stream.

18. Evidently, tools and markings that are useful (useless) at one time and place can come to be, or cease to be, useless (useful) at some other time or place. Thus we find that many objects are today judged to be art-objects which, in different times and places, were judged to be useful tools or markings. Thus, the boundary shifts with time and place.

19. Some of what is known as "criticism" and "reviewing" of art is related to placing this boundary in relation to art-objects, or art-objects in relation to it. Because every art-object is unique, each one can be the subject of such placing, that is, of criticism and review. (As paragraphs (25) to (30) show, criticism includes another important function.)

20. Especially in the case of markings, but still quite generally, "useless" art-objects, precisely because they are "useless," can be joined to other objects without altering the essential natures of such objects. When this is done, we say that the object to which the art-object is joined is "decorated" by it.

21. When the members of a particular group share a certain decoration we say that the group has adopted a certain "fashion."

22. Since decorations do not alter the essential natures of the entities that carry them, they have little or no significance for survival, and so can change from time to time, and place to place, without significant effects on evolution.

23. An extended case of (20) is that in which the art-object is joined to a playal as a program for its behavior. The art-object then becomes the basis of a secondary process, which we call a "performance," and which is not an intrinsic part of the art-object itself. In a performance, the playal becomes a "performer" or "player," and such

a performance is a member of a class known as the "performing arts." But the "useless" art-object must, evidently, when joined to the performer, not change his of her essential nature, or it would cease to be an art-object, and the performance would no longer be a "play" as performed by "players." Thus, the same "unaltered" performer must be able to perform according to the same art-object over and over again, in what can become a curious variation on the very origin of the playal, in play.

24. A performer behaving according to some particular art-object is analogous to a generalized tool running according to a certain program of markings. Strangely enough, though, a programmed generalized tool would seem to be closer to the nature of an art-object than a programmed, performing playal. This raises some questions regarding the performing non-composer and non-writer as "artist." It also suggests some interesting possible relations between music synthesizers and orchestras, as well as between puppet shows and live performances in the general context of art.

25. Viewed from the point of view of language, an art-object can be seen as having no particular *meaning*, since, being beyond usefulness, it will not bring on any *particular* behavior in a playal that experiences it. This is attested to by the frequent statements by artists that they create art-objects "for their own satisfaction," by which they intend to convey that the behavior of the playal that experiences the art object is not defined by the art-object itself. This lack of particular meaning is not to be confused with the fact that the behavior (performance) of a player can be programmed by an art-object, as in (23) and (24), since the player does not, in principle, experience the art-object, it is the playal that experiences *the performance* that experiences the art-object.

26. Since, according to (25), art-objects are without particular meaning, they can serve as "undefined signals," and can be the subject of purely individual interpretation, associated with arbitrary behavior on the parts of those who experience them. Almost all the work in art criticism can be viewed as devoted to the attempt to assign particular *meaning* to art-objects, that is, to defining behavior that could (should!) be associated with the experiencing of any particular art-object.

27. Since the process of assigning particular meaning to an art-object is one that involves assigning a possible behavior to the playal that experiences the art object, this amounts to defining a possible *understanding* of the art-object.

28. The process of definition in (27) must, evidently, include animation and narration on the part of the critic, but implies that the art-object can be viewed as an "actor" in a long narration generated by the artist.

29. We can see, from (26) (27) and (28), that the activity of the art critic is almost complementary to that of the artist, since the artist is the creator of signals that lack particular meaning, while the critic is continually engaged in trying to assign meaning. So the critic can be viewed as rescuing the art-object from its utter uselessness, by animating it, and giving it meaning, albeit the particular meaning conceived by each individual critic!

30. There would seem to be nothing that distinguishes a critic of art from any other playal that can experience art-objects and assign meaning. Thus we can, each of us, perform as critic, assigning our own meaning, as does any critic.

31. It would appear that pre-human creatures, lacking the capacity to animate and narrate, must have been limited to experiencing art at the level of its lack of meaning, that is, must have lacked the power of criticism. It would also seem that at least some schizophrenics would share this limitation.

32. There is a link between individual creatures in various states of enforced disconnection from the replicating, evolutionary stream, and art-objects. This suggests an underlying evolutionary connection between art galleries, museums, and similar kinds of repositories for objects that, one way or another, have come to "lack meaning," on the one hand, and "homes" for "old people," morgues, tombs, and cemeteries, on the other.

33. It is possible that, in some cultures, the artist is seen as achieving a significant, partial condensing out into the art-object, thus "escaping" the total destruction of death.

So what does all this have to say about "useless" art? Well, it seems to me that art is no more useless than a beach, or a river-bank, or a waterfall, or any other un-joining between one grand domain and some other. Of course, we can spend a whole life without ever going to a beach, or a river-bank, just as we can live without ever going to one of the tall galleries where evolution tumbles back to earth, sometimes clad in the silence of frozen lines and textured colors, sometimes re-singing and re-dancing as she falls. And, not going, we would miss the curious challenge that ever awaits those who travel out to that shifting edge where the storyteller can meet and bring life to the artist's useless dead, all waiting, in their quiet stillness, to be animated.

THE FARMER

The Inu are the remarkable people who make their home in the mainly frozen, icy wilderness that is Canada's north. Until very recently, the richly inventive Inuit lived exclusively by hunting and fishing, and by gathering a relatively small amount of not very abundant vegetable material. So the genius of the Inuit sustained, until recently, one of the demonstrations of the ability of humans to live without farming, that is, by hunting and gathering alone.

Not all of humanity had so long and so dazzling a passage through the sole dependence on hunting and gathering. But all of humanity must have lived in this way for nearly a hundred thousand years, before some of our ancestors made attempts at horticulture. This happened no more than ten thousand years, or about six hundred generations, ago.

The primary evolutionary significance of horticulture is clear enough, since it serves to make available more food at less risk. However, since it took something like thirty thousand generations of human existence to make its appearance, one might surmise that improving on the natural productivity of the earth must have been comparatively difficult to achieve. Evidently, no real improvement could be achieved until the organized planting of seeds and other cultivable portions of mature plants had begun. But begin it did, and horticulture became one of the interacting parts of all permanently settled communities, strengthening and strengthened by them, all at once.

The emergence of settled communities based on more developed forms of horticulture, which we refer to as agriculture, has been associated with the major patterns of modern human development. This appears to have begun in what is now Iraq, but more specifically in the smaller region known as Mesopotamia, that lies between the rivers Tigris and Euphrates, which rise in southern Turkey, and flow into the Persian Gulf.

But there is another aspect of this Mesopotamian invention that it is important to notice, for agriculture is the first activity in which the prophecy of storytelling humans is expressed directly in the hereditary development of other species. It is here, in agriculture, that, for the first time, humans begin to make *living tools* out of plants and other animals, rather than simply *use* them as tools, in their "natural" states, in the way that is as old as self-serving evolution itself. And it is here that "natural" comes to mean "in the absence of intervention by the human creature"—that seemingly "non-natural," "super-natural," "non-animal" creature.

So agriculture marks the beginning of a fundamental shift from a process of "natural" selection—based on interactions between natural genetic mutations, natural instinctive and copy-learned behavior, and natural environments alone—to a process that includes *human intervention and behavior*, with its narration, animation, and cascaded prophecy. And as surely as the making of inert tools affected the pre-human hand and brain, so did the making of living tools affect the human brain, human Isaiahs, and the nature of their prophecy.

Thus, unlike Art, that makes tools which lie at the very boundary of the evolutionary world, Agriculture makes tools that live well within it. Art is the condensation of the evolutionary past into non-replicating, "useless" tools; Agriculture is the condensation of the evolutionary past into replicating, "useful" tools. Art waits for ever; Agriculture waits for seasons. Art must still life; Agriculture must instill it. Art repeats; Agriculture reproduces. Art is canvas; Agriculture is cotton. Art mummifies; Agriculture unwraps. Art seeks vaults; Agriculture seeks sky. Art buries; Agriculture plants. Art has completions and wakes; Agriculture has harvests and festivals. The Artist cultures sterile species, and prophesies the certainties of extinction; the Farmer cultures fertile species, and prophesies the uncertainties of survival. The Artist seeks everlasting life in the freeze of death; the Farmer finds recurring life in the thaw of resurrection.

THE SCIENTIST

The occupation of scientist emerged even more recently than that of writer, or artist, or farmer, being not more than about four hundred years old in a clearly identifiable form. During these roughly twenty generations, the contribution of scientists to the ability of humans to survive has been dominated by their ability to increase the depth and breadth of understanding in humans. The efforts of the scientist yield "science," which consists of a large body of specialized narrations concerned with understanding.

To see how science has come to make such remarkable contributions to what we understand, and to begin exploring what the nature of this contribution might really be, it is convenient to recall that the play-incidents that first led to copy-learning in the playal were *episodic*, and that this led to the copies of behavior that make up understanding being themselves episodic.

And since episodes make up such a large part of what I shall be saying further about science, it is useful to notice that they have the following features:

1. A *beginning* situation and an *ending* situation that we might identify as BG and EG respectively;

2. In addition to a beginning and an ending, as in (1), internal *ordering*, such that it will always be clear the order in which situations arising, between the beginning and ending, follow each other;

3. Isolation from the other necessarily related activities which are always unfolding around the episode, in the environment in which it is proceeding.

I shall discuss two aspects of the foregoing list of features of episodes, the first very briefly, and the other at much greater length. The first aspect concerns all the features taken together, and relates to the evident resemblance between them and those of *a line on a map* from a beginning-point BG to an ending-point EG. You will see a little later on how useful this observation turns out to be.

The second aspect, mentioned in (1) above, relates to the unavoidable association of episodes with *beginnings* and *endings*, and hence to the fundamental part that these play in the nature of understanding and explanation. This is evident in the persistence with which we break up the ongoing process of existence and experience into segments with beginnings and endings, and ultimately into "isolated" events, even though our actual experience is of the continuous dissolving of the present into the next present. For example, we "start" work in the morning with some "event," and "finish" with some other "event," even though we could not identify either the start or the finish if pressed. We "are born" at some "time," and "die" at some other "time," even though what it means to be born or to die defy careful explanation.

Interestingly enough, we are aware of, and can even speak of the continuous nature of our environment and our involvement with it, as I have just been doing, but it is only with the greatest of difficulty that we understand our environment in this way, or explain it. Indeed, it is only when we see clearly that the whole structure in which we understand and explain is closely tied to the *structure of the episode*, with its *beginnings and endings*, that it becomes possible to appreciate the peculiarly unique significance of "understanding" and "explanation" based on *evolutionary* behavior, for it is in the nature of evolutionary "understanding" to be free of a tie to any beginning, and to express a continuous, endless dissolving of one state into another.

It is interesting to pursue further the beginning-ending nature of episodes, and the embedment of understanding in it, for if we combine this with the isolation that is always associated with episodes, as mentioned in (3) above, we can begin to see the origin of the presence in understanding, and especially in understanding as we find it in science, of what we refer to as a "cause" and an "effect."

To see how this comes to be, we can begin by noticing that the link between a cause and its effect is only identifiable if the continuum of situations that leads from the one to the other is so isolated that we can neglect the flow of influence, from the general environment in which the continuum is situated, into the isolated stream of situations that makes up the continuum that leads us from cause to effect. If this kind of isolation did not prevail, the influence of the general environment would introduce intervening "causes" between the beginning cause and its effect, and so deny the status of cause to the beginning of the continuum, and so on, endlessly. This therefore requires a continuum that links a cause and its effect to be totally isolated from any impinging environment. But such an isolated continuum would be beyond the reach of any kind of biological experience, and so could never be associated with actual behavior.

We can see from this that, so far as causes and their effects become features of understanding, the required isolation into episodes detached from their environments must be provided by *the process of understanding itself*, as it unfolds inside the brain of the playal, rather than by the environment of the playal which yields its outer stream, and out of which the episodes are extracted. It is very nearly this kind of isolation that the process of copy-learning provides, and which traces its origin to the intense instinctive isolation characteristic of the defensive situation of primal play.

If we now couple the beginning-ending feature of episodes to the isolation with which copy-learning and understanding also endow them, we can see why understanding is so readily expressed in terms of cause and effect, for, within the isolation that the playal brain imposes on an episode, *the beginning can be the cause of the ending, which becomes its effect*.

It is important to notice, then, that causes and effects are manifestations of the way in which the playal brain is constrained to treat the behavior that it copy-learns, that is, are manifestations of the way in which a brain *understands*. A corollary to this is that *causes and effects are not features of the environments of a brain, but are artifacts, occurring within a playal brain, of the way in which evolution has ordained that such a brain and its environment shall interact*.

But even though we can see that causes and effects are artifacts of the playal brain at work, it is interesting to ask whether there might be some continuums in the environment of the playal that are naturally isolated and organized into segments to such an extent as to *suggest* episodic behavior, and the existence of causes and effects, actually embedded in the environment itself. As it happens, there are such continuums, and they have played an important part in the work of the scientist, as well as in the life of the narrating human more generally.

It is not surprising that this should be so, since those continuums in the environment that tend to display episodic behavior will be particularly appropriate for copying by a brain that itself transforms experience into episodes in order to copy the associated behavior and come to understand it. Thus, the playal brain will be able to understand, with particular ease, those continuums in its environment that present themselves in a form that corresponds to that in which the brain itself is constrained to work. Examples of quasi-episodic continuums of this type are the rising and setting of the sun with the associated night and day, sleeping and waking, birth and death.

I should add one more strongly episodic continuum that is embedded in the environment, namely the attack by an animal-eating parent on its reared young, which is at the origin of primal play, and which has the particular distinction of having initiated the process of copy-learning, and itself set the pattern of episodic structuring that marks all subsequent copy-learning and understanding. As emphasized earlier, the necessary isolation of the attacking parent comes from deep instinctive processes in the brain of the young playal, associated with its self-interest and defense. It is this unique relation between an animal-eating parent and its carefully reared young that provides the isolation for the first of all episodes to support copy-learning; and in such episodes the beginning is always crystal clear to a threatened brain, and the ending too, except when blurred by voracious dispatch into oblivion.

I shall take a few paragraphs here to invent a word that will help to summarize some of what has just been said, and to simplify writing some of what is to follow. A word is needed that will allow me to distinguish between the pure episodes that can be fashioned in a brain, and the continuums in the environment of the brain that suggest the same qualities of isolation and beginning-ending, but which can never have the same completeness either of isolation or boundedness as the episodes in a brain, if they are ever to be able to interact with one. So I am going to keep the word "episode" for what goes on in the environment, since this corresponds to its general usage, and invent the word "episoid" for the creation in the brain. Hence, episodes are to be found outside brains, and episoids inside them.

By way of example, then, the attack by a parent on its young is an episode, and the copy-learned version of the parent's offensive behavior, which constitutes the young playal's understood version of the attack, is an episoid. As a corollary, an episoid will have a sequential quality that allows it to control behavior, and, as indicated in Chapter III, is an "abstraction," in that it is "abstracted" from an episode, and has lost all of the peripheral, concrete features of the actor in the episode, except for its behavior.

Earlier, I referred to the resemblance between the features of an episode and those associated with a line on a map from a beginning-point BG, to an ending-point EG. To derive an episoid from such an episode, imagine that the line between BG and EG is isolated by removing everything close by, on either side of the line, but with everything that it crosses left as marks accompanied by little legends on the line, so as to produce an "edited" line. All that is left is the line between B and E with markings along its length, and nothing else. This isolated, "edited" line, bounded by BG and EG, resembles an episoid.

Evidently, we can have any number of episoids between BG and EG, as well as numerous other pairs of points on the map, other than BG and EG, between which other episodes and derived episoids run, with some episoids crossing others. Also, between any two points, such as BG and EG, there can be not only as many episoids as we like, but there will be some *shortest* episoid for getting from beginning to end, that is, some episoid with the least "editing." I shall refer to this shortest episoid as "Occam's" episoid, for a reason that will soon be explained.

Returning now to my main story, when a playal copy-learns some part of the behavior in an episode, by trapping it is an episoid, the playal then *understands* such behavior in a way that reflects *the development of its brain*. In our brain (with the exception of the evolutionary case, which I shall get back to later), the episoid is bracketed between a cause and an effect, and we can also come to notice that, for some episoids, the same cause always leads to the same effect, that is, the understanding of one of a group of episodes is the same as that for all the episodes that begin in the same way. When we come on such a group of episodes, which can be understood with a single episoid, that is, whose behavior can be copied with a single routine, we say that we have come on a "law of nature," about which I shall have more to say later. However, it is important to notice that the "law," as abstracted in the episoid, is *always more isolated and more sharply bounded than any episode could be and still become known to a brain.*

We should also notice that, going back to episoids as resembling isolated paths on maps, where we have one episoid we generally can have a large number that have the same beginning and ending, that is, which link the same cause and effect. Of these, scientists generally choose the "shortest" one, which I have called "Occam's episoid," to represent the associated "law of nature," in order to add uniqueness, and what scientists like to pronounce as "elegance," to the laws that make up the narrations of science. This follows a suggestion often attributed to the fourteenth-century Franciscan philosopher William of Occam (sometimes spelled Ockham).

So what is called a "law of nature" is only a law of "nature" to the extent that a brain is a part of nature, since the isolated purity of episoids in general, and the uniqueness of Occam's episoid in particular, which together give it the character of "a law," applicable to a seemingly endless series of episodes, is present only in the brain of the playal.

Thus, although "a law of nature" must be directing us toward some underlying similarity between one episode and another, the similarity between episodes can never approach that of the identity of the single, abstract episoid of Occam derived from them, which identity supports the belief in "a law of nature." And so "laws of nature" must be, as with causes and effects, artifacts of the way in which evolution has ordained that the brain of the storytelling playal shall work, and, more particularly, of the way in which it achieves what it experiences as understanding.

Clearly, this is not to deny the evident regularities in the workings of parts of our environment, but is rather to suggest that these regularities need not be taken to reflect an "organization of nature" according to the "laws" compounded even in the brains of scientists, which represent little more than generalized carriers for "causes" and their "effects," as required by nothing more basic than the way in which we understand.

Viewed in this way, a "law of nature" can be seen as pointing directly to a recurring feature of our environment, without itself embodying the essence of the feature, not any more than a finger pointing to a fire or a river must be burning in the one case, or dripping water in the other.

Of course, even though one can come to see that "laws of nature," rather than being actual roads and highways in the environment of the brain, constitute a framework of signposts built up by understanding—which reflect, as much as anything else, simply the peculiarities of the playal brain in one of its extreme forms—it is the case, nonetheless, that the underlying regularities in the episodes, to which the "laws of nature" point, allow such a brain to achieve "almost veridical" prophecies, as I shall explain shortly.

Further, such "almost veridical" prophecies, even limited as they will be seen to be, yield great benefits to humans. And it is these benefits that justify a significant part of the work of scientists, which is the culture of laws of nature, that is, the identifying of those species of episodes for which it suffices to have a single abstract episoid in order to copy-learn, with increasing fidelity, some components of the behavior of its member episodes, that is, to understand them as a group.

When the scientist not only identifies such a single, widely applicable episoid, but explains, in a narrative, the behavior that is trapped in it, we say that the scientist has made a "theory" of the episodes to which the single episoid applies. Thus, a theory is a narrative about how the actors in a group of episodes behave, as captured by the copy-learning of their common behavior in a single, abstracting episoid. A theory will, evidently, be as abstract as the episoid of which it is the narrative, since it is the explanation of what is understood by the episoid.

Also, since theories are the narrated explanations of the behaviors trapped in episoids, they will have the same wide range of "lengths" as their episoids, and can be subjected to the constraints of Occam's suggestion, as is indeed the case in all of science—among all possible theories, the "shortest" is regarded as having particular status as the statement of "a law of nature."

Returning now to the "almost veridical" prophecies just mentioned, I shall begin by saying more fully what they are. "Almost veridical" prophecies are prophecies that are subject to what we generally refer to as "assumptions," which serve to limit the environment in which the prophecies are expected to be veridical. The assumptions limit the prophecy in a way analogous to that in which episoids and their theories are isolated and limited; that is, the assumptions serve to preclude at least certain interactions between the prophecy and the concrete complexities of our actual environment. Thus, "almost veridical" prophecies are always conditional, which is to say that they are expected to be veridical only if the conditions of the environment turn out to match those of the assumptions that accompany the prophecy, and so form part of it.

An example of such an "almost veridical" prophecy would be: "This bridge will last at least fifty years." But this could turn out to be the case, that is, this will constitute a veridical prophecy only to the extent that the assumptions about the environment, which underlie the theory of the bridge, are met, such as the load it will carry, the intensity of earthquakes that will occur, the absence of collisions with large meteorites, and so on. Analogous assumptions surround even the seemingly veridical prophecy: "The sun will rise tomorrow." In this way, assumptions allow us to substitute expressed or, as in the case of the bridge, unexpressed constraints on the environment for what might otherwise prove to be simple untruths in prophecies, thus yielding what I am calling "almost veridical" prophecies, that is, prophecies that are "veridical" *subject to* the expressed or unexpressed assumptions.

Viewed uncritically, such conditional prophecies might appear to "predict the future." But they are always less than veridical, being subject to assumptions, not necessarily expressed explicitly, but always present, as a constant witness that, while the continuing products of evolution, such as frogs and ourselves, behave, necessarily, so it seems, as though there must be a next present that dissolves out of the now present, they can never know, in the sense of a fully veridical prophecy, exposed to all the complexities and possibilities of the real environment, without conditions, what that next present will be.

It might seem strange that a substitution of assumptions for the full complexities of the environment would bring survival benefits to humans, but this arises in at least two ways. The first way in which benefits arise is linked to the fact that the "laws of nature" accumulated by humans generally, and by scientists in particular, allow a prophet to know what assumptions will often be satisfied by the workings of the environment itself, without the intervention of the prophet, since, even if the laws can never convey the essence of the regularities in the environment to which they point, the constant awareness of the regularities provided by the laws allows the prophecies to be chosen so as not to violate these known, even if limited, regularities. In this way, the prophet can make more "almost veridical" prophecies than if there were no indication of what the regularities in the environment have been.

But the second way in which benefits arise is more impressive, and is one of the keys to the enormous practical power of science. This stems from the fact that, since the laws provide lists of assumptions that suffice to make a prophecy "almost veridical," it becomes possible for the prophet to interact with the environment, with the aid of tools, if necessary, so as to ensure that at least some of the assumptions *are actually satisfied*. This increases enormously the number of prophecies that turn out to be almost veridical, compared to the number that turn out this way without such selective interaction with the environment, guided by the "laws of nature." Thus, although the "laws of nature," including those provided by scientists, cannot expose "the future," even to a narrating brain, they can serve to reduce the number of distinctly non-veridical prophecies that would otherwise be made by one.

Another example of how all this works would help to show what is involved. Let us start with the prophecy: "Place a seed in the earth, and we shall have food." This is clearly not a veridical prophecy, since we know that, quite often, seeds, when planted in the ground, don't even germinate; and so the prophecy will not in general correspond to what emerges in the future. But if we now appeal to some of the "laws of nature," we can state some *assumptions* that can begin to push the prophecy toward being veridical, and so we can restate the prophecy as: "Place a certain kind of seed at a certain depth in the earth at a place where it will have a certain amount of moisture and subsequently a certain amount of sunlight and a certain amount of nutrients in the earth, and we shall have food." As it happens, this second prophecy isn't veridical either, but it will correspond to what will actually take place much more often than the first.

However, we can go a step further, and use the assumptions as prescriptions for how to interact with the environment so as to satisfy them, and, using tools where necessary, actually provide the seed and what follows it with moisture, and nutrients, having picked a place for the seed which is exposed to sunlight, and planted it at the assumed depth. When we do this, the prophecy will correspond to what will take place sufficiently often that we will have what I have been calling an "almost veridical" prophecy.

Of course, as even many non-scientists know, what is referred to as an "experiment" is one of the keys to the invention of these apparently beneficial "laws of nature" that form part of science, and we can now make use of much of the foregoing to see what is so special about an "experiment," and the source of its unique significance. For one can now see that the scientist can make a prophecy under conditions such that, so far as science's laws of nature can tell, *all* the assumptions about the environment, which, if not satisfied, could render the prophecy non-veridical, are *actually satisfied*, as just explained, either by the environment itself or by the scientist using tools, or both.

To provide the necessary satisfaction of the assumptions, the scientist can either make an "observation" on some part of the environment, "sufficiently" isolated as it runs naturally in what becomes the "experiment," or can "set up" some necessarily limited portion of the environment in a particular way, with the aid of tools if necessary, so that it runs in what is another way of realizing the "experiment." In either form, the experiment is often referred to as being "controlled."

So now, if when *all* the assumptions have been satisfied, the prophecy turns out to be *non-veridical*, the scientist comes to know that at least one of the "laws" that prescribed one of the assumptions must be inadequate in some way, that is, must not be suitable for making even "almost veridical" prophecies. The particular law that will be exposed to the greatest doubt will usually be the law that is to be "tested by the experiment," that is, the law which, if the prophecy is veridical, could come to form or would continue to form the understanding of the episodes that correspond to the prophecy itself. But this is not always so, and the identification of the law that fails to support the scientist's prophecy might need other experiments. The process of "experiment" is continued for as long as the offending "law" remains unidentified.

And, in this way, we come on the aspect of science that is as powerful as it is unique, for what distinguishes science, from all the other narrations that follow even the careful observation of the environment by non-scientists, is the fact that, when an *experiment* indicates the existence of even a single "law" that leads to non-veridical prophecies after the relevant assumptions have been satisfied, the law is identified as false (untruth) to all scientists, at least in principle, and is no longer treated by any of them as a "law" if it had previously been, and not accepted as one if it was newly being advanced as such.

Thus, the aggregate of the laws of nature available to prophets, from science, have the unique distinction that there is not even one law, or group of laws in the aggregate, that has led to non-veridical prophecies when all the assumptions, which can be derived from the laws, are satisfied. Thus, although the laws of nature advanced by science will never be such as to support the claim of being known "always" to lead to a veridical prophecy, they do support the claim of there not having been even one prophecy which, when the assumptions of the laws were satisfied, was non-veridical, that is, of there being no law, currently accepted as a part of science, that has ever been falsified.

We can see, in this way, that the scientist is not only an active culturist of laws of nature, but also subjects them to a form of "natural selection," as the continuing streams of laws manifest their own kinds of mutations, while experiment and falsification drive some of them to extinction, with only those laws surviving that can withstand their rapidly changing environments, inside the narrating playal brain. It is this relentless and rigorous selection of the episoidal strands that evolve in the dark skulls of science, which distinguishes them from all the other means that humans employ to support their prophets.

It is important to see this—the practical and *approximate* way in which science achieves its main distinction through "the experiment"—in order to avoid an overestimate of its significance as this relates to "understanding" our environment. But it is also important to see what is the source of its *practical* power, and the way in which it drives us up against the impenetrable barrier of the future, by a relentless accumulation of more and more laws of "nature," and more and more ways of interacting with our environment, so as to make veridical prophecy look *almost* generally achievable, so rarely now, in the case of some very special kinds of prophecy, do we fail to achieve "veridicality."

Now, since this book, and others like it, contain "experiments," but might not, at first sight, appear to, I shall take a few paragraphs to explain how this comes to be. As described above, an experiment follows the decision to expose some law to falsification. But it is quite possible to subject a law to falsification by making use of an experiment that preceded even the conception of the law. When this is done, we have the same two possibilities as described before, in that the experiment could have run "naturally," as a sufficiently isolated part of the environment, or it could have been "set up" by means of specific human intervention. Regardless of whether the "preceding" experiment was "natural" or "set up," there exists the same necessity to ensure that the assumptions, on which the "prophecy" that underlies the experiment is based, were adequately satisfied by either the natural or the "set-up" conditions that surrounded the experiment.

It is important, then, to recognize that "preceding" experiments, as they occur repeatedly in this book, have just as much importance and power for testing laws as "following" experiments, and that "natural," "preceding" experiments, of the kind that form nearly all the basis for the falsification of laws in evolutionary science, are not to be dismissed as "merely finding reasons for what has already happened," since they have exactly the same relation to the falsification of laws as do experiments, whether natural of "set up," that *follow* the conception of laws.

Returning for a moment to the four-hundred year age of science, to which I referred earlier, you might have wondered how such an apparently ancient and general human activity as recording the regularities of the environment could come to be assigned such a relatively recent beginning. The reason for this is related to what has just been said about the significance of "the experiment" for the falsification of "laws of nature," since, although humans have been observing the regularities of their environment for at least a hundred thousand years, it is only within about the last four hundred that "the experiment," with all its controls, has been taken as the determinant of whether to accept or reject some law, and just within this time that only a law that has survived the test of an experiment has been taken as part of the special body of narrations that make up what we refer to as "science."

It is important to notice, then, that there is no practical list of assumptions that we could write down, even using all of science's laws of nature, which, when satisfied, would render a prophecy veridical in principle; that is, *we cannot (and will not ever), in general, formulate a veridical prophecy; all we can do with these laws of nature is reduce the number of occasions on which prophecies will turn out to be not even almost veridical.* But we should not dismiss this "all we can do" too lightly, for, clearly, it has had a very great deal to do with how we have come to the state of "mastery" of our environment we have achieved. What we need to guard against, though, is being dazzled by the enormous *practical* power of science, which flows out of its evolving, continually un-naturally selected "laws of nature," and their applications, into believing that it can ever tell us what our environment is "really like," or what "the future" is going to be.

This is a convenient place to return to the subject of myths, for we can now see that science's "laws of nature" are nothing but its myths. What distinguishes them from the more ancient myths of humanity is the filtering (un-natural selection) of the myths of science provided by experiments, and the commitment to the rejection of any particular myth if (when) it is (ultimately) falsified. So the myths of science have the purely *practical* advantage that, at any particular time, they support almost-veridical prophecy with the power that comes from *experiment*. But they do not differ from other myths in any more fundamental way than this, since, nothwithstanding our present ability to demonstrate the lack of predictive power of ancient myths, their evolutionary emergence and existence was based on such predictive power as they actually displayed. And, indeed, within science itself, we find its more ancient myths replaced by its more recent, based on demonstrations of increased capacity to support almost-veridical prophecy.

All myths, including those of science, are attempts to map the behavior of phenomena onto episoids in the narrating brain, and all myths will display the same type of "falsehood" that must exist between any original and its "copy." Further, all myths are based on the behavior of some actor that allows the behavior of the phenomenon to be animated, and the use by science of a single actor in "nature" does not distinguish its myths from those of different origins that appeal to a "single god." As we can see, then, myths inevitably provide the links between a narrating brain and the phenomena in its environment, but also limit, in a fundamental way, what such a brain can ever come to "understand" of this environment.

The foregoing discussion has been devoted mainly to the consequences for science of the constraints that flow from the episodic origin and nature of understanding. The most direct expression of these constraints is to be found in the segmented form that understanding takes, and the dominant role that causes and effects come to play in the formulation of the theorems of science. But, in spite of this, causes and effects do not constitute the only format in which science can be cast, as shown previously in the references to evolutionary science, in which the explanations rely relatively less on causes and effects, and more on the fundamentally continuous behavior of the environment, made evident through the pre-playal processes in the human brain, and which, to judge by the behavior of frogs, must also be compulsively "evident" to them.

So what I shall do now is explore, more generally, the varieties of science that can be seen to flow from the extent to which causes and effects tend to dominate the way in which the behavior of the environment is captured in the episoids that form the basis of copied behavior, and hence of understanding, and the theorems that are its explanation.

Starting with the episodic nature of understanding and the presence of causes and effects to which this leads, we find science evolving along four basic paths that yield types of science based on the following:

1. Causes with effects;

2. No causes and no effects;

3. Effects with no causes;

4. Causes with no effects.

Since so much of the foregoing discussion has been devoted to the science of type (1), it will suffice here simply to give some examples of what is intended. In this type of science we are introduced to: forces that "cause" masses to change their state of motion as their effect; a germ that "causes" disease as its effect; a drug that "causes" a headache to disappear as its effect; childhood experience that "causes" adult behavior as its effect, and so on.

The closest we come to science of type (2) is evolutionary science, in which developments precede and succeed developments, but causes and effects play a less dominant role. This is made possible by a number of features that characterize this type of science, but the "law" of natural selection is especially interesting. Indeed, the law of natural selection is hardly a law at all, and it is possible to see more clearly the source of this difference in the following way. As pointed out previously, the basis of a law of nature is the isolation from the environment provided by its episoid. However, the essential feature of natural selection is the necessity to carry the entire environment along as a part of the process in which the lives of living entities must be "understood." Thus, it does not allow the isolation that supports the identification of episodes and the extraction of their episoids, and so does not lead to understanding as can be expressed in laws with their causes and effects.

One might therefore wonder whether the principle of "natural selection" leads to "understanding" at all. The conclusion would seem to be that it does *not* lead to "understanding"—not as the more recent parts of the playal brain execute understanding—and that, paradoxically enough, this is the basis of its enormous power to reveal the general developmental process of biological entities. But if this is so, it must be that the continuous, non-episodic structure of evolutionary experience, when expressed in the form of narration, reflects a particularly powerful feature of the playal brain, in the complex form present in us, which allows it to express, in a form of linguistic behavior, experience that is proceeding at the pre-playal level. However, such experience, in its primal form, could not be structured as "understanding," since the pre-playal brain cannot support copy-learning, and so does not experience understanding. Thus, the traces of "understanding" in the narrations of evolutionary experience must come from the necessities of language and narration *themselves*, which require a particular kind of structured packaging to become linguistic behavior at all.

In this way we can see that the narrating brain rescues evolutionary experience from forever being submerged beyond the reach of spoken expression, that is, rescues it from being totally, and everlastingly ineffable. But the same brain exacts a price for the rescue, to the extent that it must package the flows of experience for narration, if only very weakly, as though they had originated in the portion of it with which the power of narration is associated, that is, as the very weakest of episodes, with *almost* no beginnings and no endings. This is not such an unusual bargain in the commerce of nature, as witnessed in the behavior of our eyes, for instance, which must break up an essentially continuous environment into discontinuous pieces so as to allow the movement of the resulting images around the insides of our brains. Or, even simpler to observe, is the lined structure of the images of television that serve an analogous purpose, outside our brains.

All this raises an interesting question as to the status of a "theory" of evolution, which is, more than anything else, a theory of natural selection, since a theory must rely on episoids that trace isolated paths through the environment, while natural selection must rely on the continuous impact of the full environment on living entities. Thus, a theory of evolution will always be plagued with a paradox, since it will always be trying to cover a map by drawing lines that have no width. That is, to the extent that a "theory of evolution" is a theory at all, its narratives will always fail to include as much of the environment as evolutionary experience itself would suggest they need; and, to the extent that a "theory of evolution" is about evolution, it will always have to be about the concerted drift of the environment as a whole, which is about more than the structure of any theory will ever fully support.

This should certainly not be cause for rejoicing among creationists, for these comments do not in any way deny the status of evolution as the basic movement in biological nature, and maybe even in all existence. All they do is highlight the limitation of even an, as of now, extreme evolutionary form, such as ourselves, in being able to express what we experience at the pre-playal level, with fidelity, at the level at which we understand, and at which we narrate and construct theories. In a curious but fundamental way, the narrating brain must *mis*represent the evolutionary world in order to convey such "understanding" of it as a brain so constituted can ever come to have.

But, much more important than merely the passing need to discourage creationists, is the fact that we can now see, in a different context, the problem of veridical prophecy. Since, to the extent that the future is a continuous, evolutionary extension of the present, it proceeds along a front so wide as to be beyond the full reach of any brain, whereas prophecy, being an expression of understanding and theory, to prove veridical, must run along a vanishingly narrow strip of this future, into which strip will always leak, notwithstanding the assumptions of the prophet, all manner of disturbance that will place a prophecy at risk, in the unending evolutionary fullness of the environment in which a prophecy must ultimately encounter its fate: outside the dark, isolating shelter of a fertile, maternal brain.

And in this way we can also see the source of the power of an evolutionary science that makes no room for species that would survive by looking into an unknowable future, and which entertains only those whose members live in real nows, built from real thens, that have no next nows beyond those reached by their own survival in the fickle little pieces of the total, present environment, in which they must continue riding, if they are to ride at all, to any other nows.

I shall move on, now, to a discussion of science of type (3), which, as mentioned above, is characterized by "effects without causes." This is a sufficiently strange-sounding basis for science that it warrants careful introduction. So I shall begin by calling attention to the fact that most humans, even now, are limited in their experience to what they can derive from their own biological endowment. Thus, most humans, even today, have never observed their environment through a microscope, or a telescope, for instance, and so, except for the second-hand process of narration, have, as the basis of their continuing contact with their environment, the senses with which they are endowed.

We can see from this that, notwithstanding the presence of tools which extend the faculties of some present-day humans, it must be the case that our brains are the evolutionary products of a relation with our environment that did not include tools such as microscopes and telescopes which extend our ability to see. Thus, to the extent that our brains can produce theories which have causes and effects, these must be a reflection of the nature of a brain for which experience was limited to that derivable from our normal biological endowment.

However, to the extent that understanding, and especially the filtered understanding typical of science, is limited to causes and effects, there will arise, in the pursuit of such understanding, "effects" for which the "causes" reside below the level of our normal biological capacity to observe. Thus, if we are to place such effects within the framework of understanding and theories, it becomes necessary to *invent* a "cause" for the "effect."

Using a microscope, for instance, we can pursue the cause below the level of our natural endowment, and "see" the "cause," as with a germ, but even this process eventually exhausts itself, and we are left finally with "effects" that we can "see," if only with a microscope, but for which the "causes" remain beyond even our augmented capacity to see. Such "effects," for which the "causes" are beyond the reach of even our extended ability to achieve "direct" experience, I refer to as "*effects without causes,*" since the evolutionary basis of causes and effects is linked to direct experience; and so, "effects," which lack such direct evidence of what could be taken to be causes, are, in an important evolutionary sense, without causes.

In the presence of such "effects without causes," scientists *invent* causes, since the structure of understanding is such that "every effect *must* have a cause." Of course, the causes are not invented in isolation, but are attached to theories, which serve to link the invented cause to the effect which, prior to the invention, had no cause.

This invention of causes, known only through their effects, has become the primary work of the part of science known as "physics," and especially is it the work of the part of modern physics that is concerned with what scientists refer to as "fundamental particles." The behavior of these "particles," including the forces that determine the relationships between them, is explained by what is known as "quantum mechanics."

What I shall do now is show that the nature of the "fundamental particles" invented by physicists, to serve as causes, seems to be strongly affected by the structure of the *grammar* that forms the basis of narration, and which, in turn, must form the basis of any theory that could tie an effect to a phantom cause. Stated differently, I shall show that some of what is regarded as the proper content of physics, and of quantum mechanics in particular, seems to be as much an expression of the constraints of grammar as of anything happening in or between the "fundamental particles" themselves.

If you are unfamiliar with the language and process of quantum mechanics, it is worth noting that what you are about to read is not part of the standard fare in that demanding discipline, but rather an attempt to demonstrate that some of its most basic statements can be read so nearly as simply statements about grammar, that the stamp of what must be used for narrating, on what the narration is supposed to be about, seems almost inescapable.

This need not be surprising, since it could be that, when pushed beyond the evolutionary limits expressed in its biology, as is the case with "effects" that have "causes" which are beyond even the augmented senses, and which can no longer be shared by acts like pointing, the human brain might be unable to separate distinctly the actors from the framework in which they must be situated in narration, that is, in any theory. This raises a question about where the scientists who explore the physics of "fundamental particles" might be leading themselves, and us with them; but I leave this for comment at the end of the demonstration that follows.

We can begin the demonstration by seeking a cause that is "those 'things' that must be down in there, and their relations." And since "those 'things' that must be down in there, and their relations" are a cause, this must be the beginning part of a narration, which is to say that it must be the beginning sentence in a narration, which, in Occam's sense, must be the "shortest" theory that explains the connection between "those 'things' that must be down in there, and their relations," on the one hand, and the effect, on the other. Thus, the limiting "those 'things' that must be down in there, and their relations" that the quantum-mechanics narration will convey are the parts of a sentence:

$$Q\ B\ L,$$

and its five purely structural passives

$$Q\ L\ B,\ B\ Q\ L,\ B\ L\ Q,\ L\ Q\ B,\ L\ B\ Q.$$

To the extent that the causal "those 'things' that must be down in there, and their relations" of quantum mechanics are connected to "fundamental particles," those that it will put into a grammatical story, which can form a theory, are then Q, B and L.

So, what might the quantum mechanics story say? What, based now just on grammar, might we expect a "theory of fundamental particles" to be like? We might expect that the theory will sound much like the following:

1. There are two kinds of fundamental particle in the world. One kind – the Q, L kind – always occurs in pairs, while the second kind – the B kind – serves as a sort of glue-particle, to hold the other two together. These form a basic kind of much bigger, dissociable particle, which is the natural form of the world, in which Q, B, L are together as a whole. This QBL particle deserves a name, say "F" particle. Seen as grammar, Q and L are "words" situated as subject and object with respect to B, another word, which is a verb; and F is a sentence.

2. The only "explainable" version of the world consists of a pair such as QBL -- LBQ that is symmetric, and pairs such as QBL – LQB, of which there are four, that are anti-symmetric. This kind of pairing is evidently related to the possible active-passive structure of the most primitive form of language.

3. A typical quantum-mechanical way of saying what we can see to be the case in paragraph (2) would be that the world is made up of symmetrical and anti-symmetrical pairings distributed with a frequency (probability) of one-fifth for the symmetrical and four-fifths for the anti-symmetrical arrangement.

4. The "binding force" produced by the particle B is the most basic of all bonds, and without it the structure of the world cannot be understood, and would simply disintegrate. The related grammatical statement would be that the only way to construct a sentence is for the verb to "bind" the subject and object together, with what is, evidently, the ultimate grammatical "binding force," without which a sentence cannot be understood.

5. From the QBL particles – the F particles – we can build whole hierarchies of bigger and bigger "particles." That is, sentences can be strung together to form whole hierarchies in the form of narratives of increasing length.

6. If we are to have strings of *sentences*, that is, of F particles, which connect up to make a story, then each sentence will have to have one on its right that is *different* from the one on its left. Thus, each F particle can only be involved in a unique way, or, stated differently, no two F particles can occupy the same state. Also, since the fundamental particles are in the Q state and the L state with respect to B, they also cannot occupy the same state. This has the sound of what is know in quantum mechanics as "Pauli's exclusion principle."

7. Evidently, any "effect" that the particles of quantum mechanics might have will be through their changes of relations, that is, as a result of changes in how the words and sentences are assembled into stories. However, due to the discrete nature of the words and sentences, the "effects" will always be *discontinuous*, because there are no intermediate states for words, which must be in one definite place or another; that is, there is no intermediate state for the particles. When the sentences are few, the effects will be very clearly discontinuous, that is, "quantized." When the sentences are very numerous, that is, with large aggregations of F particles, the effects will tend to seem continuous, because of the very large number of possible states, and hence the relatively small effect that any single change of a word will make in general. In this way, the quantum world of invented causes

in the form of "particles" finally merges with the continuous world of evolutionary experience and behavior.

8.   There is nothing that emerges from the foregoing to suggest that "those things that must be down in there" of quantum mechanics need be "particles," in the sense of being little spheres which are simply too small to be seen. Indeed, it is well known that "packets of waves" can also serve as "those things that must be down in there" of quantum mechanics, that is, as its causes, and satisfy the effects to which they have to be joined.

9.   As the ability of the scientist to "'see' down in there" in greater detail increases, there would be yet another level of "effects" which would be "identifiable" below the level of Q, B, and L. These effects would then have "causes" that would have no "meaning," because the "causes" would have moved below the level at which grammar supports meaning, having shifted to the level of word-fragments, that is, of morphemes. Since, for a small number of words, such as Q, B, and L, the number of morphemes is larger than the number of words, it would seem possible that below the level of the particles Q, B and L, the particle-physicists are engaged in the identification of a growing number of meaningless fragments which will never form part of the grammar of narration, and therefore will increasingly be beyond being understood and being inserted into any theory.

Although the last nine paragraphs are hardly an introduction to the rigors of quantum mechanics, they begin to raise the question as to whether, at the limits of the interactions with our environment that we are able to induce, physics determines, without limit, what our biology will understand, or whether our biology, as manifested in the fundamental limitations of grammar, for instance, determines what even the most elaborately instrumented physics can ever reveal as theory, and as part of what is understood.

This is a question that begins to have considerable practical importance, since, as if to emphasize it, the cost of the monster installations needed for further exploration of "fundamental particles" is now such as to involve even the wealthiest of countries in expenditures that are huge on the scale of even their enormous resources. (All this was written some twenty years ago—before the latest and biggest had been built.) It would therefore be reassuring, for those who must support such expenditures, to hear from particle-physicists why they believe that they are not now increasingly involved in the most expensive investigation of one of the more uninteresting parts of grammar ever devised by the playal brain, partly trapped, as it might be, in a ceaseless search for glimpses of *itself*, in hidden causes.

Science of type (4) is in need of explanation at least as much as that of type (3), relating, as it does, to "causes without effects." There is only one instance of such science of which I am aware, and this is the science of what are known as "black holes." But even this one instance is instructive, for black holes find their theoretical basis in Einstein's general theory of relativity, and so they tell something about this comparatively recent and powerful part of science.

So what is a "black hole," and how does it come to be a "cause without effect"? A "black hole" is an enormously massive "thing," located for away in the space of the cosmos, which has the capacity, by virtue of its massiveness, to attract into itself every kind of particle, and which does this with such ultimate and definitive force as to prevent any evidence of its existence from escaping, by attracting and holding to itself any particles that would emerge from it and provide such evidence. Thus, a black hole can "cause" the particles that come within its attractive force to disappear into itself, without revealing what has become of them, that is, without "effect." A different way of presenting this strange behavior is to say that the black hole is a situation in which there is no "effect" distinct from the "cause" itself, that is, the black hole is an instance of the limiting case in which a "cause" is its own "effect."

From the point of view of grammar, the two ways of presenting a black hole are both extremes. The first way, that of the cause without effect, can be viewed as a narrative theory that is so long that it takes an extremely large amount of "events" and editing, that is of time, to get from the cause to the effect; so long in fact, that the narrative never arrives at its end. According to this view, the black hole will eventually, if one waits "long enough," reveal itself in effects, but there is no way to know what "long enough" might turn out to be. The second way, that of the cause which is its own effect, could be the manifestation of a theory consisting of a single sentence, so that the beginning and the ending lie on top of one another, with no editing, and no need for time. According to this view, the black hole has done all the revealing it is going to do, and there is no other "effect" worth waiting for. In both cases, the black hole seems to be accompanied by a virtual destruction of the significance of "time."

You might have noticed that some of this sounds not unlike what we were just discussing in particle physics, and so one might wonder whether there should be any connection. I believe there should be, and that the connection comes from the distinct possibility that, what we are looking at in the case of the black hole, as the limiting "particle" of *cosmology*, is similar to that in the case of the limiting "particles" of *physics*, in that the behavior of these portions of our "environment," about which theories are being constructed in the both extremes of science, at its micro and macro ends, is so far from any possibility of direct experience, of the kind that accompanied developments in the playal brain as recent even as narration, that the narrative theories are now loaded with the structure of narration *itself*, at least as much as with the "actors" that increasingly escape distinct separation from the matrix in which more directly experienced actors were, and still are distinctly conveyed.

What the physics of "fundamental" particles and cosmology share, more assuredly than anything else, is the narrating playal brain. And so relativity and quantum mechanics tend to lead us toward the same kind of extremity of the capacity of this brain to support understanding; one of them exhausts itself in the immensity of the cosmos, and the other in the innermost smallness of things.

In view of this, I repeat, in a slightly modified form, the question raised earlier as to whether, at the frontiers of what we now seem to be able to come to "experience," physics and cosmology determine, without limit, what our biology will understand, or whether our biology, as manifested in the fundamental limitations of understanding and grammar, determines what even the most elaborately instrumented physics and cosmology can ultimately reveal as theory, and as part of what we understand.

I believe that biology imposes limits, not only on what science can ultimately tell us about our environment, but on how science can tell it, especially as this concerns the edges of that environment, as we must come to "experience" them through instrumental surrogates, whether the edges be too far out, too far in, or too far back for "direct" experience. Hemmed in by these limits, the only route by which science could add anything to the experiencing of our environment, as distinct from just deeper understanding of vanishingly thin strips of it, is via the paradoxical partnership between narration, with its episodic necessities (and which owes nothing to science), at one level of our development, and, at another level, continuous evolutionary experience, confirmed by experiments (and this owes as much to the lawfulness of science, as to its laws).

And this is what makes the evolutionary stream of science so absolutely singular in its pointing to the distant and continuing links of creatures in general, and humans in particular, to their environment. For, in seeing, even dimly, *when* and *how* we came to be, we can also glimpse *what* we came to be; and hence fix, if only roughly, where we end and where our environment begins; fix what is part of our passing environment and what is just a passing part of us. And maybe that's the most the scientist can ever help us do, and why Charles Darwin will always seem so exceptional in having helped us do it.

Interestingly enough, we are now in a position to predict what has become one of the most notable features of modern science. I am referring to the tendency, which is most evident in cosmology, for the explanations of science to oscillate between two basic forms: one based on beginnings and endings, and another based on continuous evolution. We would, on the basis of the sort of arguments I have been making since Chapter III, but especially those in this Chapter, expect this to be so as an expression of the continuing and completely inevitable difference between the *episodic nature of understanding* as it arises in the playal brain, and the *experience of existence as continuous and evolutionary* as it arises in the pre-playal segments of the *same* playal brain. The nature of this difference is such that, to the extent that the episodic nature of playal understanding must gain expression through grammar, such expression can never match the underlying experience of existence as essentially continuous and evolutionary; while, to the extent that the continuous and evolutionary nature of existence gains expression, it can never be satisfactorily conveyed by the grammatical language to which science is limited for its explanations.

And so the explanations of science will always oscillate between the modes of explanation that express these two irreconcilable features of the playal brain. This is, evidently, not a feature of science alone, but must pervade all our attempts to explain our existence and the nature of our environment. But the irreconcilable nature of the underlying features of the playal brain becomes a particular challenge and problem for science because of its insistence on the rejection of the falsifiable as a part of its narrations, since the one of these features of the playal brain will always serve to falsify the expressions of the other. In this way, the very basis of the practical power of science becomes the driver that sets it oscillating near the limits of its explanatory power, as the grammar in which we are caged allows us to see, between its bars, a universe that has no other marks than the shadows of our cage.

A recent and dazzling instance of this inevitable problem for science is beautifully documented in *A Brief History Of Time* by the brilliant physicist-cosmologist-philosopher Stephen W. Hawking. Indeed, a half-cycle of the oscillation is documented as occurring not just in "science" in an abstract way, but in the life and work of a great contemporary scientist, that is, in a single narrating brain. This is exactly what would be expected as the purest case of the aspect of science to which I have been directing attention. The oscillation occurs between support for a "big bang" version of the "beginning" of the universe, and an evolutionary version that has no "beginning." What is clear is that cosmologists have no way of bringing this oscillation to an end, since it reflects such a fundamental mode of behavior of the narrating playal brain, combined with the commitment that science has adopted to rejection of the falsifiable.

This paragraphs was written in 2009, not 1988 as are almost all the others. It is an extension of the ideas in the preceding paragraph, driven by the present existence of the Hubble telescope, which was set in orbit only in 1990. But the truly remarkable Hubble, for all its ability to see "far out" into the universe, is really just a way of extending what our eyes can see. And it is important to recall that eyes, much like ours, were already present in the heads, next to the brains of pre-playal reptiles. So, what our Hubble eyes reveal will always be what pre-playal eyes reveal, that is, an evolutionary version of the world, free of beginnings and endings. As a consequence, no matter how "far out" our Hubble eyes will ever see, they will never see where the universe begins or ends. And it is only in the narrating brain that the "understanding" of what they see will, of necessity, speak of an un-seeable beginning—a big-bang or other beginning—which resides, among the host of other episoids, with their beginnings and endings, *nowhere but in the narrating playal brain itself.*

You might have noticed that the first occasion on which I have used the word "time," in anything like a technical sense, was about ten paragraphs back, in connection with black holes. The reason for this is that "time" presents such difficulties in discussion that I have avoided it in the hope that, the more background I accumulated, the easier the discussion would become. It seems that whatever gains are to be had in this unsure way must have been realized by now, and so, irresistibly, the time(!) has come . . . And, alas, the gains seem not to have cancelled the difficulties altogether. The first indication of this will be my starting the discussion of "time" without even an attempt at a definition, relying on "the general impression of time" to carry the introductory paragraphs until a fuller discussion of the origin of what we refer to as "time" allows me to go back and situate the introductory usage in a fuller context.

As a place to start, this also has a more general kind of appropriateness, since "time" has played such an important part in what scientists have had to say. But, as with some other subjects, we do better to start with a frog's view of it, than with a scientist's. And, indeed, the frog has a view, for we can notice that, when it ejects its tongue at a fly, it doesn't wait from sunrise to sunset to get it out and back. What happens is that it not only does this in a quick flash, but it does so in a way that corresponds to what it would have to do to catch the fly, which, evidently, has its own interest in moving and saving itself, and so won't take from sunrise to sunset to do this either. It would, however, be more instructive if the frog *did* take from sunrise to sunset, since then we would see easily that, while its tongue is going out and back, other things are happening as well, and not just in the fly. What the "flash" tends to hide is the fact that there has never been a frog such that its tongue can go out and back so that nothing *else* will happen while this is happening. This "something-else-that-is-always-happening" is the first suggestion of "time," even though all that matters to the frog is that the "something else" shouldn't include the fly's departure.

What we can see here, as in the case of prophecy, is that the frog has an instinctive mode of behavior that suggests "time," the "time" in which its tongue must move in order to catch the fly. But, at least as interesting is the fact that the "time" of the frog's movement is linked to that of the fly's. And so it must be that, not only do the frog and the fly have an instinctive way of suggesting "time," but it is "similar time," or there would be either far fewer frogs or flies.

Of course, we could extend the story of capture, or some similar story, to frogs and other creatures, as well as to flies and other creatures, and, in this way, extend the instinctive suggestion of "time" to include all creatures. In the case of ourselves, we need not look far to see that we also embody such instinctive "time," for we move with instinctive motion, as frogs move, to brush away a fly, as it approaches one of our eyes, for instance, unbidden.

So it seems we can conclude that evolution has led to the embedment of a common suggestion of "similar instinctive time" in the genes of all creatures. But this is just another manifestation of the evolutionary relation of all creatures to their total environment as it smoothly dissolves; and "time" expresses the fact that this goes on endlessly, everywhere that experience has known.

By proceeding in this way, I have illustrated at least one of the difficulties that beset the discussion of "time," since, in one paragraph, I talk about an episode in which the frog's tongue goes out and back, as though this separates out cleanly from all the other activities that are going on, then, in the next, I talk about a continuous environment, from which such episodic activity could be extracted only as part of what is understood inside the playal brain.

In such a smooth, dissolving environment, there can be no "events" to "mark" the beginnings and endings of "periods" of "time"; hence, for the frog, "time" must proceed without "markings," and be nothing more than that "something else that is always happening." But, in us, the situation is different, for, in addition to sharing the frog's instinctive "time," we speak of a "time" that comes from clocks, complete with marks, all ready to be cut up. But how could all this come to be; how could the "time" of the frog have evolved into the time that we use when we speak of the "time" to fly from Ottawa to Vancouver, or the time for a pulse of light to travel from Sun to Earth?

We can begin an answer by noticing that biology shares with the inanimate not just space and matter, but "time" as well, for a great mountain can no more "live" outside "time" than a great ape can. Indeed, by far the most compelling evidence of time and its passing resides in the rocks of mountains, and their sands. And so it is no surprise that the first scientists to wrestle with the link between "time" and evolutionary experience were geologists, among whom Charles Darwin should properly be numbered. This raises the immediate question of what the biological basis of such "time" might be.

It would seem that the existence of instinctive "time" must come from the chemical processes and substances that underlie all of biology, pushed mainly by energy from the sun; and I shall come back to this shortly. The tendency for "time" to be related, from one species to another, must come from the incessant interactions of various species and the general influence of natural selection.

The first stage in which "time" is other than the awareness that "something else is always happening" must have developed with the emergence of copy-learning in the playal. The principal factors that would begin to isolate "time" are the episodic nature of copy-learning, with its beginning and ending "events," and with its corollaries of understanding and selfness. It is here that the episoid makes its appearance, and begins to lay the biological groundwork for an isolated stream of the kind required by the final isolation of "time."

But although the episoid that stores copy-learned behavior is an ordered sequence, that is, has a record of what follows what, it does not carry the same degree of genetic control as instinctive "time," and this is evidenced by the extent to which copy-learned behavioral routines can be executed in a "time" different from the instinctive behavior on which they are based. So that episoids begin to support a less rigid "time," as just another part of less rigid behavior generally. Far from being a disadvantage, this allows the copy-learned routine to be executed according to the detailed circumstances in which the playal in question finds itself, its behavior being relieved of the necessities of a pre-determined, genetic "time." To the extent that each playal applies copy-learned behavior according to its own particular circumstances, the combination of instinctive "time" and copy-learned "time" serves its needs well.

However, the situation changes much further with narration, where it becomes necessary to insert, directly into the narrations, references to events in the environment that have sufficiently *common, shared* features, in order to make the narrations suitable for linking behavior in a number of playals, regardless of the separation between them. Without the insertion of such shared events into narratives, they would lack the capacity to determine the behavior of the hearer with useful precision, and would have no capacity to contribute to survival. Without such references, the narratives would not be transferable from one playal to another, and so would fail to develop the most powerful features of narratives, namely the ability for prophecies to undergo cascaded modification from one playal to the next.

For example, it is not very useful to say: "John and his family have not eaten." Much more useful is "John and his family have not eaten since the river flooded their camp." It is the insertion of an event, "the river flooded their camp," an event that is sufficiently *common*, that allows the hearer of the narration to place the "have not eaten" within some behavioral framework that can lead to beneficial, related action. This need to insert common events does not exist in discourse where the grammatical first and second person suffice, and where both the speaker and the hearer are immersed in almost the *same* "something else that is always happening," that is, in the same "time."

It is the occurrence of the inserted common event in a narration, among all its other events, that becomes the reference for "when," in "time," the other events occur; and it is the separation between inserted common events, the "duration" between them, as this occurs in the environment naturally, that becomes the basis of some sort of common "interval of time." And it is this "natural separation between inserted common events" that becomes the "time" which accompanies narration, that is, which becomes understood by us most generally as time—the time of history. Thus, the inserted common events in a narration serve to prevent the other events that make it up from falling on one another, that is, they hold the other events apart, even while holding them together.

But there remains a truly grand question about "time" that needs answering, and it has to do with the role that "time" plays in science, and in the science of moving masses in particular, which is generally referred to by scientists as "mechanics." For, without "time," the laws of the mechanics of Newton simply could not have been formulated, and without "time" the even more stupendous mechanics represented by the relativity of Einstein would have no basis. "Time" is not a bystander in science, and especially not in the mechanics of either Newton or Einstein. It is an active player that forms an integral part of laws and prophecies; and the prophecies are remarkable in the frequency with which they are almost veridical. Everything that "happens" in mechanics happens in "the time" of the experiment, or the phenomenon, or the problem.

So the grand question to which I referred is one that is so simple that it is almost impossible to pose; is so simple that it is almost a mystery: How could event-related "time," which is associated with nothing more basic than the relatively recent emergence of the playal brain and of narration, come to seem to be such a fundamental part of the behavior of our environment, and of the way in which humans generally, but scientists in particular, record and explain its behavior?

The answer would seem to be that the power of "time," in the narrations of science, derives from the *particular kind of naturally occurring shared events* that were inserted into the narratives that became the basis of "time" in science, and especially in mechanics. And, to emphasize how particular these were, we should notice that there is an unlimited range of events that can be inserted into narration for marking "time," since we could use events such as the eruption of a volcano, or the growing of a tree, or the birth of a person, or the flooding of a river, and so on, endlessly.

However, there is a particular series of events, the series of *positions of the sun*, which, when inserted into narratives, constituted the insertion of not just *any* shared events for marking "time," but, tacitly, of a *statement* about *the way in which a body moves*, that is, a statement that a body moves "uniformly" when "un-hindered." For the succession of events represented by "positions of the sun," especially as they are seen from a place near the equator, are simply an expression of the uniform motion of the mass of the earth as it moves (rotates), mainly unhindered, around its axis.

Thus, what might appear, at first sight, as "just another way of inserting events" into narration, which leads to day, hour, minute, and so on, is really the insertion of a *statement*—albeit tacitly—*about the behavior of a part of our environment*. That is, the insertion of "time" into narrations, by the use of events consisting of successive positions of the sun, constitutes the insertion, into narration, of a "regularity of nature" having to do with *how "un-hindered" masses move*. Hence, what we tend to regard as "just" the use of "regular" or "uniform" "time," in such narration, constitutes the repeated use of one of our so-called "laws of nature" having to do with the way in which "un-hindered" masses move.

So that when Newton and Einstein appear to be able to embed the particular one of the innumerable possible "times" of narration, based on "positions of the sun," in laws about the behavior of masses, they have really done something quite different in the invention of their mechanics.

What they have done is use the motion of one mass—the mass of the earth— to describe the motion of all masses; that is, they have constructed a "mechanics of masses" about the behavior of masses generally, using the behavior of one such mass, carried along as the "time." This brings mechanics, as it involves the use of "uniform time," much closer to being a *tautology*, than any sort of mystery about how "time" could come to play such an important role in the explanations of the behavior of masses in our environment.

We should not be disturbed at the appearance of tautology, since its repetitions assure consistency, albeit at the expense of enormous difficulty in saying anything that is not apparent in the initial statement on which the tautology is based. But it is precisely the overcoming of this difficulty that marks the brilliance of a Newton or an Einstein.

But all this says something about how extremely fundamental must be the feature of the environment to which we point with an "un-hindered" body moving "uniformly." For this is all that "time," as used in science, would seem to be about, and so it must be at the root of all mechanics, and especially relativity mechanics, which takes a very sensitive position with respect to "time." Indeed, looked at in this way, we can see that the inclusion of "time" in mechanics is a way of carrying along, during all the narrations that make it up, the fundamental belief that an "un-hindered" body will move "uniformly," so as to make it impossible for the narrative to neglect this aspect of the way in which we understand our environment.

We can run the foregoing discussion in the other direction and conclude that all ways of "keeping time"—time of the kind that can be used in mechanics—must reduce to observing the un-hindered motions of masses. This is evident in the case the balance-wheel of what is now an old-fashioned watch, and not so evident in the case of a pendulum. And, although I shall not explain the details, this is also the case in the oscillators (vibrators) that employ wafers of quartz to "keep time" in not so old-fashioned clocks and watches; and it is so in the atoms of elements such as caesium which can be made to execute remarkably un-hindered motions and so can be made to "keep time" with remarkable "accuracy." So, to "keep time," something must "keep mass," and continue to perform an "observable," "un-hindered" motion.

This also begins to suggest why relativistic mechanics comes closer to a basis for understanding the environment than does Newtonian mechanics; since, if, as it seems to be, the "time" of mechanics is just a restatement of the behavior of un-hindered masses in space, the complete separation of time and space in Newtonian mechanics must point somewhat away from what is going on in the environment; while the coupling of time and space, into space-time, as in relativistic mechanics, must point much closer to the goings on in our environment that lead us to speak of "time."

We can also see why, if relativistic mechanics arrives at a black hole, the "time" in the black hole should be drastically affected, since, if we are cut off from any possibility of experiencing the motions going on in the black hole, or if there are "no motions left" in it, we are also cut off from its "time." It is only if we fail to realize that "time" in mechanics is a *tautologous restatement of the motion of masses*, that we can be surprised that masses and their motions can affect the "flow" of "time."

Of course, "un-hindered" bodies exist only on episoids, and so any conclusions that scientists arrive at, based on the use of this abstract version of motion, must eventually reflect the differences between the environment itself, and the unreal version of it, no matter how useful for understanding, that resides exclusively inside the playal brain. All of this stresses the unavoidably approximate nature of "time," and there are unavoidable consequences that accompany its approximate nature.

Perhaps the most common of these consequences is associated with the notion of "future," since "uniform motion" would seem to allow one to "get into the future" along the episoid that carries "time." However, not only will that "future" always be limited to the vanishingly narrow strip of experience that characterizes the episoids that carry "un-hindered motion," and so not be able to represent the fullness of the real future, but it will be limited by the abstraction on which "uniform time" is based, namely the abstraction of an un-hindered body in motion, and the necessary approximation that accompanies this. So we can also view this inability to "get into the future" as a manifestation of there being nothing *approximate* about the next present into which "now" dissolves; the actual future is *exactly* what and when it turns out to be, and so the approximations that always beset "time" will always close the door on the future as approached along "time."

But there is another whole side to "time," and we can come on this by noticing that *masses and their images* do not move either together or in the same way. Clearly this is so, otherwise it would not be possible to see an object from many directions at once, since, if the object moved with its image, it would have to move in many directions at once. Even simpler is the fact that we do not have to be moving with an object to see it, and so its image must be moving in a different way from the motion of the object itself. What this says is that whatever mechanics scientists use to describe the motion of masses should not be suitable for describing the movement of their *images*, since the two kinds of motion are apparently different.

And, indeed, there is a whole "mechanics" of the motion of images, which is different from the mechanics of masses, and which happens to be called "electrodynamics." This electrodynamics was assembled by the famous scientist James Clerk Maxwell from the results of many experiments made by himself and others, as well as a particularly brilliant insight of his own. One of Maxwell's more interesting conclusions was that the images of objects all travel at a quite definite speed, which can be measured in an experiment, and which can be taken as the "speed of light," and will include the speed of all "electromagnetic" waves extending from radio waves to all the light we can see and beyond, past x-rays and even further beyond.

This is not the place to say much more about the details of this mechanics of images, which is also known as the "electromagnetic" theory of Maxwell, but it is important to notice that "time" plays an important part in this theory of how images move. And since there are no masses moving in Maxwell's mechanics, only images, one wonders if a "time" derived from the movement of masses would work in his theory. As it happened, Maxwell used the same "time" as that used by Newton for his mechanics of moving masses, and succeeded in assembling a theory that describes the movement of images with remarkable accuracy.

This should certainly raise doubts about my assertion that the "time" of science is simply a restatement of the way un-hindered masses move, for, if this were so, it would be surprising that this kind of "time" could serve to describe the motion of images separated from masses. Indeed, if my assertion is to survive Maxwell's demonstration, it must be that there is a link between how masses move and how their images move, which is not carried in Maxwell's theory, or it would not be possible to use the same kind of "time" in his theory.

The simplest way of rescuing my assertion would be to see if, notwithstanding the remarkable success of Maxwell's massless theory of images, the images of masses also *do* have mass, and so his theory of the motion of images could use the same "time" as that which serves Newton's theory of the motion of masses. A refinement of this would say that the masses of the images of objects must be very small compared to the masses of the objects themselves, or Maxwell's theory shouldn't work at all, since it neglects the masses of images altogether.

As it happens, this is exactly the case, but it took the brain of an Albert Einstein to show not only that (in effect) an image has a mass, but how to calculate its magnitude. Of course, masses pull on one another in a way that we refer to as "gravity"—the larger the masses and the closer they are to each other the greater the pull—so it should be that masses *also* pull on *images*, if these have mass.

One way of doing an experiment to test all this is to see if the light coming to us from a distant star, that is, the star's image, when passing close to the mass of the sun, is pulled toward it and deflected from the path it follows when not passing so close. This was one of the first experiments in which Einstein's theory of relativity was tested, and it showed that images *do* behave as though they have a mass, thus, incidentally, rescuing my assertion that the "time" used in science to describe the motion of masses is the basis of a tautology; or, that such "time" is itself simply a restatement of the un-hindered motion of some mass, the earliest to serve this purpose having been that of the earth. (If one's assertions must be rescued, good that this be done by a Darwin or an Einstein!)

But now that the assertion has had such a prestigious rescue, it is quite fitting that it be allowed to go on the "offensive" and assert further that only those entities that have mass admit of "time"-based descriptions of their motions of the kind that are found in the mechanics of (Newton or) Einstein, for only then will the tautology of mass-motion "time" apply. What this says is that, if there are entities in our environment that have no mass, then descriptions of their motions using the "time" of mechanics, that is, the "time" we get from clocks of whatever kind, should not be possible, since, to the extent that such entities change position at all, their motions are not simply special cases of the motion of masses, that is, are not cases to which "time" relates as in mechanics.

This is not an entirely trivial speculation, since scientists, some years ago, identified a "particle," named by them the "neutrino," which is considered as possibly having zero mass. So what the assertion is saying about the neutrino is that, if it does have zero mass, the description of its motion will need some as yet non-existent non-"time"-based mechanics.

It is interesting, and, I believe, significant that when we pursue the consequences of nothing more basic than an outgrowth of the narrating playal brain, we should come on what appears to be a fundamental problem of physics.

We can now go back to the beginning of this discussion of time, and surmise why it is that there can be anything such as an instinctive suggestion of time that can be seen as being within the general framework of time. This would seem to be due to the fact that, to the extent that biological processes depend on chemical processes, which themselves are dependent on the motions of the masses of atoms and molecules, biology must, at some level, suggest the kind of "time" that we associate with mechanics, or, at least, it must admit of being understood in this way. What we need to be sure of, though, is that this "instinctive" suggestion of "time" is our *own* suggestion, made to ourselves, using our way of seeing "time" as "going by" us.

And it helps not at all to say: If only the frog could speak, it would tell us what it knows of time. For, if it could speak, it would see "time" through our same glass, etched in cyclic seconds. Since it is the narrating brain that casts the blanks on which "time" is scribed, and takes the rare hand of a Newton, or a Maxwell, or an Einstein to mould its tautologous metaphors into the figures of understanding.

THE PRIEST

To see where, among all our occupations, fits that of priest, it is useful to notice the following situations in which myths can, at any particular time, be placed:

1. Myths that have been falsified, that is, which have been shown not to support almost-veridical prophecy;

2. Myths that cannot (yet) be falsified, that is, which have been shown to support almost-veridical prophecy;

3. Myths for which it has not been possible to devise any way of demonstrating that they are or are not falsifiable.

Myths in situation (1) are not entertained by normal humans, because they lead to untruths in prophecies; they tend to produce disadvantages with no advantages, and so are not self-serving. Myths in situation (2) are the only ones claimed by scientists, and their generation and application make up the major part of the occupation of scientists. Myths in situation (3) can be either entertained or rejected by normal humans. If they are entertained, such behavior is referred to as based on "faith" in the myth.

The occupation of priest, to the extent that it can be distinguished from that of scientist, includes those activities that flow from entertaining myths that are in situation (3), that is, includes those activities that flow from faith. An example of the difference between the two occupations comes about as follows: since there is no experimental way of falsifying the myth that relates to "life" after death, scientists do not entertain it, but some priests entertain it "on faith." Thus the narrations of some priests concern the possibilities flowing from this myth, but such possibilities are absent from the narrations of science. Since narrations can come to control behavior, the narrations of priests can come to control the behavior of groups of humans in ways that reflect the presence of this myth in their narrations.

Evidently, acts of faith are specific to individual priests, and reflect details of their pasts. However, acts of faith can lead to entertaining mutually exclusive or contradictory myths, and so it becomes necessary for certain groups of myths to be associated with one priest, and certain other groups of myths to be associated with another priest.

We can see, from this, why the scientist seems to have been continually reducing the range of myths available to expressions of faith, since, by devising new ways of falsifying certain myths, some number that could have been entertained on faith are eventually falsified, reducing the number of myths available for entertaining on "faith" alone.

But we can also see that, to the extent that science appears to displace faith, it simply replaces one group of myths with another. What it doesn't do, because it never can, is replace a myth with something *else*. And, since all myths are inevitably incomplete, *for only the phenomenon itself can ever be complete,* myths will, all of them, in time, come to be falsified, revealing the work of science, as it runs in every present, to be just acts of suspended falsification, disguised acts of faith, needing priests, and worthy of a preacher.

EPILOGUE

The words of the Preacher, the son of David, king in Jerusalem.

Vanity of vanities, saith the Preacher, vanity of vanities; all is vanity.

What profit hath a man of all his labor which he taketh under the sun?

One generation passeth away, and another generation cometh: but the earth abideth for ever. The sun also ariseth, and the sun goeth down, and hasteth to his place where he arose.

The wind goeth toward the south, and turneth about unto the north; it whirleth about continually, and the wind returneth again according to his circuits.

All the rivers run into the sea; yet the sea is not full; unto the place from whence the rivers come, thither they return again.

All things are full of labor; man cannot utter it: the eye is not satisfied with seeing, nor the ear filled with hearing.

The thing that hath been, it is that which shall be; and that which is done is that which shall be done; and there is no new thing under the sun.

Is there any thing whereof it may be said, See, this is new? it hath been already of old time, which was before us.

There is no remembrance of former things; neither shall there be any remembrance of things that are to come with those that shall come after.

I the Preacher was king over Israel in Jerusalem.

And I gave my heart to seek and search out by wisdom concerning all things that are done under heaven; this sore travail hath God given to the sons of man to be exercised therewith.

I have seen all the works that are done under the sun; and, behold, all is vanity and vexation of spirit.

That which is crooked cannot be made straight: and that which is wanting cannot be numbered.

I communed with mine own heart, saying, Lo, I am come to great estate, and have gotten more wisdom than all they that have been before me in Jerusalem: yea, my heart had great experience of wisdom and knowledge.

And I gave my heart to know wisdom, and to know madness and folly: I perceived that this also is vexation of spirit.

For in much wisdom is much grief: and he that increaseth knowledge increaseth sorrow.

ECCLESIASTES: Chapter I

REFERENCES

Bloom, Floyd E.; Lazerson, Arlyne; Hofstadter, Laura, 1985, *Brain, Mind and Behavior,* New York, W. H. Freeman and Company.

Chase, Michael; Weitzman, Elliot D.; Editors, 1983, *Sleep Disorders Basic and Clinical Research,* New York, Spectrum Publications, Inc.

Chomsky, Noam, 1957, *Syntactic Structures,* Mouton and Co., The Hague.

Clemente, Carmine D.; Purpura, Dominich P.; Mayer, Florence E.; Editors, 1972, *Sleep and the Maturing Nervous System,* New York, Academic Press.

Darwin, Charles, 1859, *The Origin of Species.* Since 1958 a reprint has been available as A Mentor Book, Times Mirror, New York.

Dawkins, Richard, 1976, *The Selfish Gene,* Oxford, Oxford University Press.

Dement, William C., 1972, *Some Must Watch While Some Must Sleep,* San Fransisco, W. H. Freeman and Company.

Dennis, M. and Whittaker, W., 1976, "Language Acquisition Following Hemidecortication," *Brain and Language* 3 (1976), pages 404-433.

Fagen, Robert, 1981, *Animal Play Behavior,* Oxford, Oxford University Press.

Guilleminault, Christian; Dement, William C.; Passouant, Pierre; Editors, 1976, *Narcolepsy,* New York, Spectrum Publication, Inc.

Hawking, Stephen W., 1988, *A BRIEF HISTORY OF TIME,* Toronto, Bantam Books.

Hewes, Gordon W., *Language In Early Hominids, 1972, Chapter One, Language Origins,* Silver Spring, Maryland, LINSTOCK PRESS, 974.

Hobson, J. Allan, 1989, *SLEEP,* New York, Scientific American Library.

Lenneberg, Eric H., 1966, *Biological Foundations of Language,* New York, John Wiley & Sons.

Luce, Gay Gaer, 1965, 1976, *Research on Sleep and Dreams,* U.S. National Institute of Mental Health, DHEW Publication No. (ADM) 76-244.

MacLean, Paul D., *Triune Brain,* Encyclopedia of **Neuroscience,** Boston, Volume II, 198?, edited by George Adelman, pages 1235-37, BIRKHAUSER.

Palmer, Frank, 1971, *Grammar,* Harmondsworth, Penguin Books Inc.

Pincus, Jonathan H.; Tucker, Gary J.; 1974, *Behavioral Neurology,* London, Oxford University Press.

Restak, Richard, 1984, *The Brain,* Toronto, Bantam Books.

Springer, Sally P. and Deutsch, Georg, 1981, *Left Brain, Right Brain (Revised Edition),* New York, W. H. Freeman and Company.

Wegener, Alfred, 1966, *The Origin of Continents and Oceans,* New York, Dover Publications. This is a reprint of the fourth edition.

Wilentz, Joan Steen, 1968, *The Senses of Man,* New York, Thomas Y. Crowell Company.

www.ingramcontent.com/pod-product-compliance
Lightning Source LLC
Chambersburg PA
CBHW032059280526
45784CB00012B/125